UNEP Studies Volume 2

Environment and Development in Africa

UNEP Studies

Volume 1 The Effects of Weapons on Ecosystems
J. P. Robinson

Volume 2 Environment and Development in Africa
Environmental Development Action (ENDA)

Volume 3 The Child in the African Environment
R. O. Ohuche and B. Otaala

UNITED NATIONS ENVIRONMENT PROGRAMME
Project No: FP/0404-78-05

A previous publication within the scope of this series was published directly by UNEP, but is available from Pergamon Press:

TECHNOLOGY, DEVELOPMENT AND THE ENVIRONMENT: A RE-APPRAISAL

A. K. N. Reddy

Environment and Development in Africa

A study prepared by

the Environmental Development Action (ENDA)
for the United Nations Environment Programme

Published for the
United Nations Environment Programme
by
PERGAMON PRESS
OXFORD · NEW YORK · TORONTO · SYDNEY · PARIS · FRANKFURT

U.K.	Pergamon Press Ltd., Headington Hill Hall, Oxford OX3 0BW, England
U.S.A.	Pergamon Press Inc., Maxwell House, Fairview Park, Elmsford, New York 10523, U.S.A.
CANADA	Pergamon Press Canada Ltd., Suite 104, 150 Consumers Rd., Willowdale, Ontario M2J 1P9, Canada
AUSTRALIA	Pergamon Press (Aust.) Pty. Ltd., P.O. Box 544, Potts Point, N.S.W. 2011, Australia
FRANCE	Pergamon Press SARL, 24 rue des Ecoles, 75240 Paris, Cedex 05, France
FEDERAL REPUBLIC OF GERMANY	Pergamon Press GmbH, 6242 Kronberg-Taunus, Hammerweg 6, Federal Republic of Germany

First edition **1981**

British Library Cataloguing in Publication Data

United Nations Environment Programme
Environment and development in Africa. - (United Nations Environment Programme. UNEP studies; vol.2).
1. Africa - Economic conditions
2. Man - Influence on nature - Africa
I. Title II. Environment Training Programme
III. Series
330.9'6'03 HC503 79-41261

ISBN 0-08-025667-8

Printed in Great Britain by A. Wheaton & Co., Ltd., Exeter

CONTENTS

LIST OF DIAGRAMS

LIST OF MAPS

LIST OF TABLES

INTRODUCTION

One may well wonder what forms development in Africa will take between now and the year 2000, and to what degree environmental considerations will be taken into account. Given the environmental problems and development options, what premises and indicators can support a valid discussion of Africa's development strategies?

Simply stating the question gives an idea of its extreme importance and the difficulty of giving even a tentative answer. Therefore, the approach we have proposed in the following pages is limited in terms of both time and the resources deployed. The goal of this glance into the future of Africa is above all to stimulate discussion. The most difficult aspect of this venture has undoubtedly been the comparison of what is necessary and what is possible in terms of basic needs and potentials in the areas of agriculture, industry and health. For a better evaluation of foreseeable trends it would have been useful to create concrete alternatives for production, consumption and human establishments: an extremely difficult task which we have only begun to undertake here.

Even venturing no farther than it does, the approach adopted here appears extremely risky, first of all, because the harmony which characterized the strategies of national independence struggles in Africa was not transformed into harmonious development strategies during the period which followed. The result has been a lack of unanimity of points of view with regard to the various aspects of the future of the environment in Africa. There is also the obstacle constituted by the various usages of the terms *environment* and *development,* as well as the many levels at which they are taken into account. For example, the environment as it appears to a peasant group is not always understood either by officials in charge of local or regional development, or by the technicians of the various departments who are to assist in planning the environment. Added to this is the multiplicity of conceptions of perspectives and forecasts.

Projects which do exist are greatly scattered, and their aims and origins have little in common. Often they imply tacit hypotheses or options which, if not clarified and examined, can, to a greater or lesser degree influence the interpretations which may be made of them. "All forecasts", writes Landing Savane, "imply that the socio-political evolution in the different countries of the region will not experience any upheavals during the period in question. But in the case of the African continent, the last quarter of the 20th century will in all likelihood be marked by such upheavals" (141). In addition this approach, which underscores serious difficulties and unsound orientations, is not likely to be welcomed with open arms. One should not be surprised if it elicits more contention than constructive comments, from both governmental representatives who have a clear vision of things concerning their own countries, and specialists who have a thorough knowledge of the facts and tendencies within their respective purviews.

It is scarcely necessary to add that the elaboration of the elements of the perspective presented here constitutes an arbitrary approach. The choice of examples, the way in which facts and tendencies are grouped together, and the preference given to certain hypotheses over others, all may be challenged. It would be useful for the purposes of discussion that other approaches be compared to this one.

This work must be kept in proper perspective. It is much more an instrument of analysis and reflection, taking into account orders of magnitude and tendencies, than a collection of precise forecasts of the future; it is more speculative than predictive.

This is a first stage: there must be more thorough-going research, by geographical zone and by speciality, in order that those who are to decide the future of Africa may be better equipped to evaluate the possible impacts of their choices.

Chapter One
CONSTRAINTS AND CHALLENGES

The elements of the history of the future which are already visible in the African reality of today need to be determined and examined. In other words, one must pinpoint the real areas of choice which exist in light of the constraints implied by the present conditions and trends in Africa, and in the face of the challenges they present.

The population that has already been born and which with each passing day grows more numerous will not simply be forced to contend with cyclones, periods of drought, the disappearance of forests, the advance of the desert or the processes of erosion and degradation of soils which are currently taking place. These people will have to feed themselves, look after their health, and exercise a socially profitable employment. The people of Africa will thus be forced to take into account the constraints and tendencies of the African environment, which will limit the possible range of scenarios which may be envisaged.

HOW MANY AFRICANS?

In the year 2000 there will be more than 800 million Africans (see map 1). The possibility of rapid population growth in most of the African countries no longer seems catastrophic if one considers that each additional mouth to feed also means two arms to work—on the condition, however, that land is available, efficient techniques disseminated, and an adequate social organization created. In fact, most of Africa is underpopulated. The real question is whether the increasing numbers will be gainfully employed or whether, on the contrary, they will swell the ranks of the underemployed and the jobless.

Several initial remarks should be made. The forecasts made by the United Nations are based on a series of speculations concerning the future which ought to be closely examined.

The birth rate of the African countries is expected to decrease progressively as generations go by. While the birth rate in Africa decreased by less than 2% between 1950-55, the hypothesis chosen by United Nations demographers predicts that the drop in this rate will exceed 7% during the last quarter of the century: by region, birth rates over this period would decrease by 11.7% in North Africa, 6.8% in southern Africa and more than 6.0% in West and East Africa. While hypotheses as to the evolution of cultural and religious values in Africa are particularly risky, it is safe to assume that attitudes which favour a decreasing birth rate will continue in the year 2000 to hold sway among most of the population of the continent, though it is possible that there will be a progressive evolution in behaviour in the large cities. As for the death rate, it would decrease as well, and the life expectancy at birth would increase by a half point each year in the average

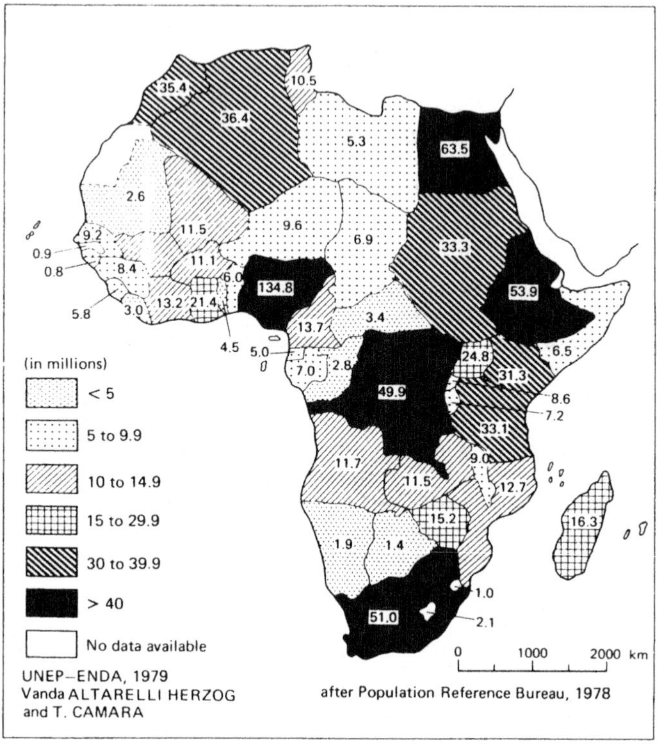

MAP 1. Population forecast—year 2000.

hypothesis. Just as for the birth rate, forecasts of a lowered death rate are "optimistic". In both cases, it seems that the United Nations experts, who worked out the forecasts which have generally been adopted for the work on the future of Africa, did not sufficiently take into account either the situation among the poorest classes where the governments have implemented a policy of birth control, or the fact that many African countries in fact fall into the category of the poorest countries. Migratory movements are only taken into account for North Africa, and not for the rest of the continent.[1]

Even with these reservations, if the average hypothesis of the United Nations is referred to, one must anticipate that the annual population growth rate will rise from 2.64% in 1970-75 to 2.88% in 1995-2000 (see map 2). These forecasts are based on hypotheses concerning the trend in birth and death rates and their components. However, the factors which determine births and deaths are extremely varied, and their evolution difficult to

[1] Despite the illustrative interest of the demographic extrapolation curves, it would be incorrect to treat population growth as an external variable. Its evolution should be foreseen in the light of socio-economic development. A slackening in demographic growth will probably take place once basic material needs have been satisfied, depending on the policies adopted in the different countries and the evolution of social attitudes.

foresee. Most recent studies reveal a great diversity in birth rates due to the variable combination of different aspects of the natural and cultural environment.[2]

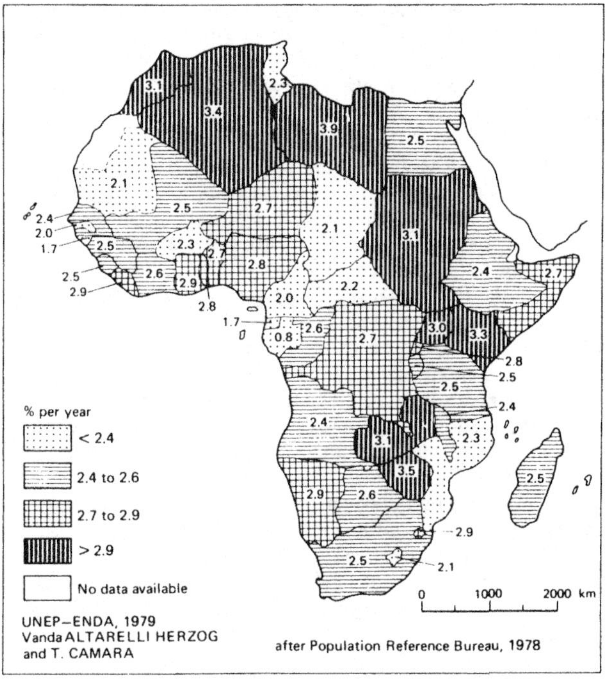

MAP 2. Natural growth rate.

In any case, it should be expected that by the year 2000 the number of young Africans between 5 and 15 years of age will exceed 200 million. The working population will have to bear the costs of supporting and educating a high proportion of children and adolescents. The relationship between age groups over the next two decades might in fact emerge as follows:

AGE GROUP RELATIONSHIPS

	0-4 (%)	5-14 (%)	15-64 (%)	65 and + (%)
Africa as a whole (1985)	17.9	26.6	52.5	3.0
Africa as a whole (1990)	16.4	26.6	53.7	3.3

It is difficult to determine the probable distribution of the African population over the various regions of the continent in the year 2000. It is likely that West and East Africa will

[2] For example, in many African cities, since births are no longer planned as they were traditionally, the birth rate is increasing.

continue to keep pace with each other demographically; according to the average hypothesis the figures would be 238 million for the former and 239.9 for the latter, while for North Africa they would be 191.8 and for central Africa, 87.7. These figures could be significantly modified by the amplitude of migrations within and without Africa.

Within most countries, the continuation of the rural exodus appears inevitable. The chances are great that the growth of the cities will continue at a rapid rate during the two coming decades without a complementary increase in activities in the cities or the creation of an adequate number of new jobs.

It is important to remember that the forecasts which have been presented here could be modified considerably by the action of governments and other forces which may influence attitudes and behaviour. This is also true to a limited extent for the size of the population, although it must be realized that the effects of a slackening in the growth rate over the next few years would only become apparent after the year 2000.

Factors which might be more responsive to change are inter-state migratory currents (we have witnessed in the recent past the massive expulsion by certain African and European countries of African migrants) and the magnitude of the exodus toward the cities. Yet one must be aware that, whatever the foreseeable political orientations may be, it will not be a simple matter to change the distribution between urban and rural areas, to equip the latter adequately and guarantee to the rural dweller both an acceptable standard of living and ongoing participation in decisions which concern him or her. The city will continue to be the place to seek one's fortune, drawing individuals with promises of emancipation and modernity.

This tendency will be difficult to reverse because the exodus has become a custom which the young people neither can, nor wish to, abandon. Whether it be in Rif or Kabylia, or in the Mossi country, among the Jerma of Niger or in the states and territories which are turned toward the Republic of South Africa, this pattern is firmly established.

This growing population will need to be fed, guaranteed satisfactory sanitary conditions, and integrated into productive economic life.

HOW THE CONSTRAINTS OF THE PHYSICAL ENVIRONMENT MAY BE LIMITED OR OVERCOME

Some of these constraints are independent of human actions. Others, however, are the direct or indirect result of human activities.

In the face of phenomena such as earthquakes or cyclones there is little that can be done. In the future, they should be taken into account in planning construction methods and in the selection of certain types of crops over others.

A number of elements support the conclusion that the destruction of the forest, in particular, and the desert's advance (which will be discussed further on) may provoke significant transformations in climates at the regional, local, or even continental level. In any case, it is known that the type, rhythm and length of the seasons, the pattern of winds and rain, insolation, and evaporation circumscribe the choices relating to crops or livestock to a certain extent.

Most specialists, even if they do not agree as to the medium- and long-term evolutions of the climate, feel that the climate is becoming increasingly variable, and that as a result, agricultural production may be more seriously compromised in the future than in the recent decades (122). The problem therefore lies in identifying the opportunities available over the coming years for controlling or compensating for these climatic vicissitudes. The refinement of forecasting, irrigation, and extension work, the selection of more resistant species of food plants, food stocks, the perfection of means of rapid transport of food products from one zone to another, and adequate distribution mechanisms, are some of the aspects which must be taken into account.[3] Given the difficulties of the present situation, the range of possible policies remains enormous.

This is also true for deforestation, the advance of the desert, and the erosion and depletion of the soil. These problems constitute, and will continue to constitute in the year 2000, undoubtedly the most serious threats to the African environment.[4]

The present trend is toward a rapid disappearance of dense forest cover, under the combined pressure of commercial exploitation and land clearing for cultivation. Commercial exploitation removes trees without renewing them, and draws the forest's inhabitants to work-sites and cities. In certain regions of Kenya, forest destruction has attained a rate of 5 to 10% annually. In the Republic of the Ivory Coast, if the present pattern of exploitation is pursued, the forest could disappear entirely in less than 15 years (18).

In the Sahel, the extrapolation of present tendencies leads to even more alarming forecasts. Wood provides the bulk of energy in an environment which depends on the preservation of a sufficient forest cover. At present, consumption of wood in the Sahel varies between 0.5 and 1m^3 *per capita* annually (34). The forecast of the trend in demand for firewood between 1975 and 2000 is dismaying:

CONSUMPTION OF FIREWOOD
(in thousand m^3)

	1975	2000
Urban	3,000	9,800
Rural	13,000	23,700
TOTAL	16,000	33,500

However, during the same 1975-2000 period, the production of wood will undoubtedly decrease as a result of accelerated deforestation. Thus, it would be necessary to plant approximately 4.5 million hectares in the Sahel between now and the year 2000 (34).

Reforestation in the Sahel represents only one-fiftieth of the effort required. In over half of the African countries the planting of trees ought to be increasing at least tenfold, if it is to have an impact over the two coming decades. Otherwise, the spread of desert-type ecosystems will accelerate. "Each year, in North Africa as a whole, the desert claims 100,000 ha of land. In the space of twenty years, the thorny steppe has advanced 90 km south of Khartum" (72) (see map 3).

Approximately 55% of Africa is threatened by the desert's advance, and 45% is subject to severe droughts (66). Map 4 indicates the risks of desertification, according to gravity. Nothing less than the preservation—or destruction—of over half of the African territory and approximately one quarter of the population of the continent is at stake.

[3] It is not very likely that other foreseeable solutions—for example, the relocation of the population away from the most endangered zones—will be adopted between now and the year 2000.

[4] These aspects have been given impressive treatment in various publications of UNEP and the FAO, particularly the World Conference on Desertification (179).

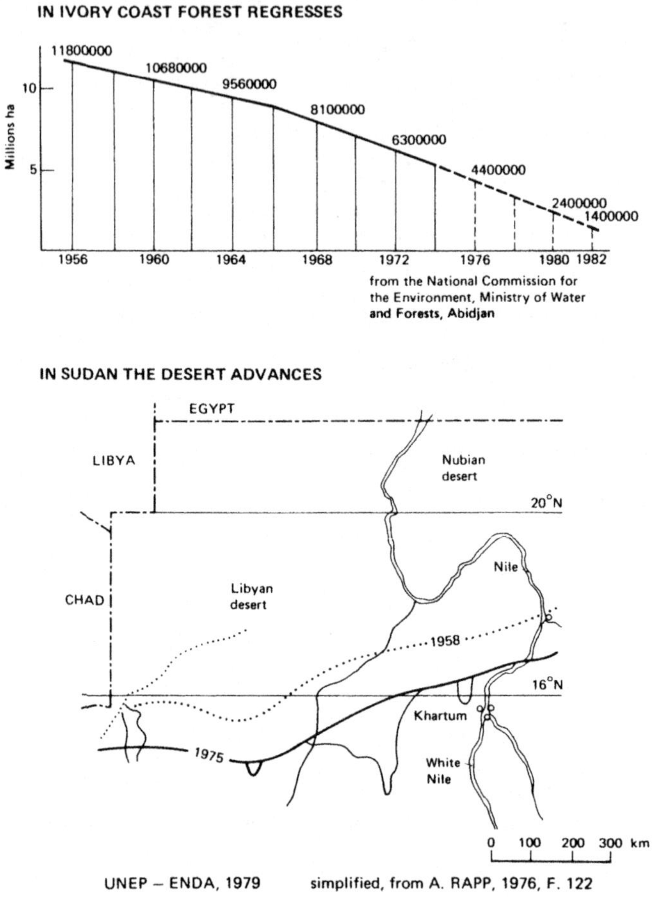

MAP 3. Tendency for the destruction of the dense forest and advance of the desert.

A "green belt" to halt the desert's advance, such as the one planted in Algeria, constitutes a foreseeable solution. Another consists of the organization of groups of herders and peasants living in the endangered periphery, who would be equipped to plan and manage their grazing areas and their landholdings. The chances of either contingency being successful will depend on the type of development scenarios adopted.

The phenomena referred to above generally provoke erosion and accentuate it. "The combined effect of the burning of land for cultivation and the activities of the forest industry is threatening to destroy the tropical forests of central Africa . . . and to result in a dangerous erosion of the soils" (179). The insufficient attention devoted to erosion is un-

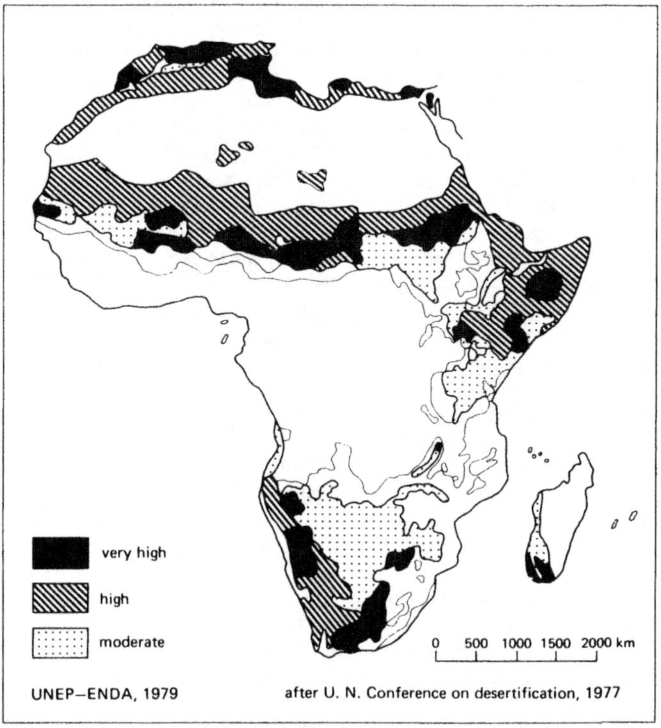

MAP 4. Risks of the desert's advance in Africa.

doubtedly due to the fact that its forms are varied, the process of deterioration which it causes are often greatly protracted, and its effects are unequally felt. Around Kilimanjaro, for example, erosion does not seem to conflict with harmonious development of the zone. In a country such as Lesotho, the process of gullying alone causes a yearly loss of arable land of 0.5% (122). In Africa as a whole, it is estimated that the annual rate of loss due to erosion is probably close to 0.25%. This destruction of the soil takes place over decades, but its effects are significant and far-reaching.

A general campaign to control erosion calls not only for a national effort for seriously affected areas, but also a long-term, broad-based campaign in which the peasants and herders as a whole are involved. There is reason to fear that in many African countries the conditions necessary for the implementation of generalized action for erosion control may not be fulfilled by the year 2000. In consequence, the damage to agriculture and livestock by the governments of the most seriously threatened countries will be much greater than imagined.

Lack or oversupply of water—the latter appears paradoxical in a continent where aridity is so common—constitutes constraints for a number of reasons. There may be total or seasonal shortages; the drought and the desert's advance have just been discussed. For perhaps a fifth of the rural dwellers in Africa, organizing themselves according to the water available, saving it and using it efficiently, have traditionally been the fundamental objectives. It will remain so for the next two decades, even if the public authorities frequently contend that this task is now up to them.

The choice of priorities in the allocation of large quantities of water, between the development of the valleys or lakes (Senegal, Niger, Zambezi, Chad), or the irrigation of many small perimeters should certainly become a focal point for discussion of pressing national problems. Another option which would call for exhaustive debate is the allocation of scarce water resources for cash crops which are expected to procure the money necessary to buy foodstuffs, or for food crops.

Within this context, the problem of the exploitation and management of underground water and possibly its replenishment, and the organization of irrigation will appear over the coming decades as special domains of the agricultural research and extension work effort in the arid and sub-arid zones.

The frequent scarcity of water should not be allowed to camouflage the difficulties which are caused when it is overabundant. This may occur in dry areas or in those areas where erosion has taken a particularly heavy toll, such as in Morocco or Mozambique. Excess water also appears in natural areas such as the mangrove or marshes, as well as in areas where it rains incessantly throughout the year. It is likely that attempts to develop areas of abundant vegetation will multiply. Their success depends on the attention given to the existing ecosystems and the danger which may result from their mismanagement.

An entire series of constraints in the African environments concerns the enemies of crops and livestock in the broad sense of the term. These are rodents, birds (the keleo-kelea for example), insects (the locust being one of the most famous) and cryptogramic diseases of viruses. The tse-tse fly is particularly significant among these pests because it creates a barrier which is practically impenetrable to bovine livestock. The amount of damage caused varies with the enemy and the zone in question but a number of economists feel that it may frequently attain something on the order of 5% of the harvests.

Various methods of pest control have been employed in Africa: traditional procedures to begin with, then the action of specialized agencies (Organisation Commune de Lutte Anti-acriolienne et de Lutte Anti-aviaire) for example. During the two coming decades decisions will be made in favour of either a chemical campaign, with the generalized propagation of modern procedures—which will undoubtedly predominate in the countries with the highest revenues or in the enclaves constituted by the large plantations—or other methods which involve less pollution and are more appropriate for use by peasants or herders.

Africa is faced with many other constraints which cannot be analysed here. One of them, which warrants individual treatment, is the limitation and unequal distribution between African countries of energy resources (see map 5). The fact that at present the exploitation of energy resources in Africa is limited may not be as serious as it was thought to be.

For coal, in addition to the countries which are currently producing and are indicated on map 5, Algeria and Tanzania also have minor production, Zaire has reserves, as does Botswana, and the Malagasy Republic has the Sakoa coal basin; this country's local needs have been negligible until now, and the fleets of the Indian Ocean prefer to use fuel oil. In the future, particularly with the rise in the price of petroleum products, there will undoubtedly be a renewed interest in the coal deposits which for the present are being neglected. The sum total of African coal reserves is estimated at 75 billion tons.

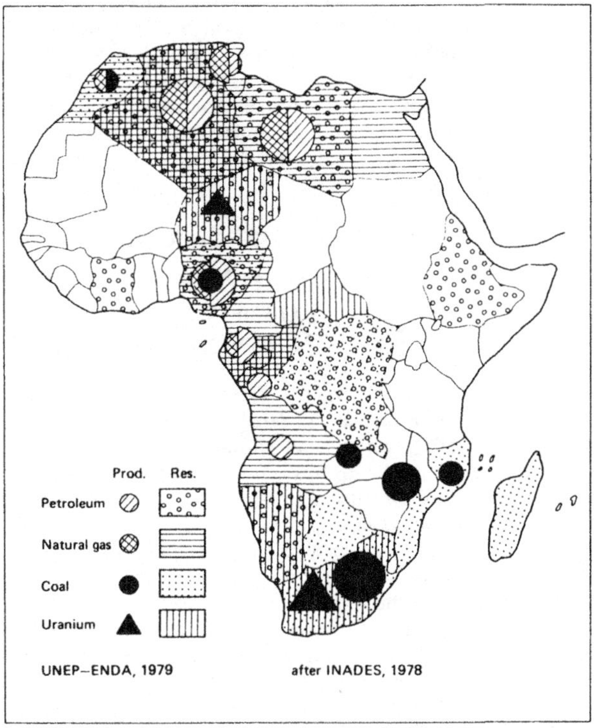

MAP 5. Present energy resources and reserves.

With regard to petroleum, in addition to the countries whose production is indicated on the map, one should include those countries in which large reserves have been identified, such as the centre-west region of Chad, the Ogaden (Ethiopia), the Republic of the Ivory Coast, and the coasts of Cameroon and Zaire. There are also serious forecasts of finding oil off the coasts of West and southern Africa. Algeria's proven natural gas reserves appear colossal: in the order of 3 billion m^3 and production forecasts would allow increased exportation to Europe and possibly to African countries. It seems possible that the plan to construct gas ducts across the Sahara will regain some of its credibility over the coming decades.

Africa at present provides one quarter of the world production of uranium. The largest reserves are found in the Central African Republic, the Republic of Gabon and Niger, while other reserves are found in the Republic of Mali, Mauritania, Somalia and Chad. Africa has large quantities of different minerals (see scenario 1 in Chapter 2).

One of the major questions of the next twenty years centres on the level of interest and effort which will be shown by peoples and governments to overcome the constraints and reduce the risks which have just been mentioned and, whenever possible, eliminate them. This raises the problem of the intensity of research, training, and all kinds of actions which might be taken, as well as the resources (in terms of credits and voluntary work) which could be mobilized to carry them out. There seems little likelihood that, between now and the year 2000, the attenuation of these constraints will constitute a specific objective of government action or a matter of daily concern to the people. There is much more chance of this type of action being carried out if it is associated with a more primary concern, that of satisfying the people's basic needs.

FINDING ANSWERS FOR BASIC NEEDS

Providing Food for Africans

If it is true that the primary need of the African people is to obtain an adequate food supply, then any study of the future should give preference in its analyses to the food and nutrition problems.

Most of the projects which have emerged are global in scale, covering either the continent as a whole or a large region such as the Sahel. A similar effort has not been made to estimate needs over the coming two decades within different environmental zones, administrative districts and village groups. Better local appraisal of needs and, simultaneously, of possibilities, constitutes an important approach which should be effectively carried out over the coming years.

It is important to begin by becoming aware of the magnitude of the food shortages which can be foreseen for Africa over the coming decades. To understand the problems it is necessary to evaluate some of the food and nutritional problems which will have to be overcome and to reflect on the scope of the solutions to be implemented.

Among the forms of dependence which have characterized Africa in the past and which will need to be shed over the coming years, food dependence constitutes a particularly disquieting problem. While the reasons underlying the growth in needs are known, sufficient stress may not have been placed on the fact that the population is rapidly increasing within a context in which on the one hand export crops are monopolizing land and other production, and on the other the urban demand is rapidly growing. Added to this is the influence of certain imported models of consumption. It is in the cities that these tendencies are most visible.

A more detailed analysis of the various components of food requirements expressed in different African countries should be made as soon as possible, so that the reliability of forecasts may be increased.

While the indicators, based on different factors and often applying to different groups of countries, paint dissimilar pictures of the food situation, they almost unanimously concur that this situation is deteriorating.

The gap in Africa between the growth rate of food production and that of the population has been assessed in different ways. According to some sources, in recent periods there has been a slight surplus:

FOOD SURPLUS

1962-65			1962-72		
Population	Food production		Population	Food production	
	Total	*Per capita*		Total	*Per capita*
2.2 %	2.2 %	—	2.5 %	2.7 %	0.2

From 155 and 120, 121 and 122.

According to other sources, for the period from 1960 to 1970 the increment in gross food production was negative (122). Using the base figure of 100 in 1961-65, the World Bank calculates that the *per capita* food production index for the African continent as a whole has evolved from 99 in 1966-70 to 96 in 1971-76 (204). And R. Brown offers the following evaluation of the exportation and importation of cereals in Africa (in millions of metric tons) (23):

IMPORT/EXPORT OF CEREALS IN AFRICA
(in metric tons × 10^6)

1934-38	1948-52	1960	1970	1976
+1	0	-2	-5	-10 (estimate)

In its examination of the situation by zone, the FAO observed a tendency toward a relatively severe food shortage in the various parts of the continent (67):

FOOD SHORTAGE BY ZONE IN AFRICA

Level of self-sufficiency	North	Sahel	West	Centre	East and South
in 1962-64	99	99	100	99	99
in 1972-74	78	83	94	94	98

What is important is that the information currently available and the extrapolations which can be made do not indicate a reversal of this tendency. Nevertheless, it should be noted that demand forecasts imply hypotheses as to purchasing power, tastes and patterns of consumption. For each of the forecasts of food requirements, the implicit or explicit hypotheses used by their authors should be known.

Forecasts of the margin between production and demand in the countries of the Maghreb suggest that, according to the average hypotheses, in 1985 there will be a shortage of 285,000 tons of meat and 1,252,000 tons of milk and dairy products. Similarly, in the Arab-speaking countries of Africa[5] a shortage of vegetable oil is expected in all the countries, except the Republic of the Sudan and Tunisia, and of raw sugar, except in the Sudan and Somalia.

According to reasonable estimates, an annual growth in the food supply of 4% would be necessary to satisfy the needs of Africa as a whole. In addition the availability of transport and adequacy of purchasing power would be necessary to distribute this production where it is most needed.

The gap between foreseeable production and needs in 1990 (if the aim is to attain an adequate nutritional level, estimated by the FAO to be 110% of standard caloric needs) is in millions of metric tons of wheat equivalent: 5.5 for Egypt, 14.6 for Nigeria, 5.7 for the continental countries of the Sahel and 10.1 for East Africa (87).

[5] Algeria, Morocco, Tunisia, Libya, Sudan, Egypt, Mauritania and Somalia.

This situation creates a demand for agricultural growth (see map 6) which will be difficult to satisfy. For many African countries, the prospects in terms of food self-sufficiency appear limited. They are even more limited if the kinds of foods and their nutritional compositions are taken into account.

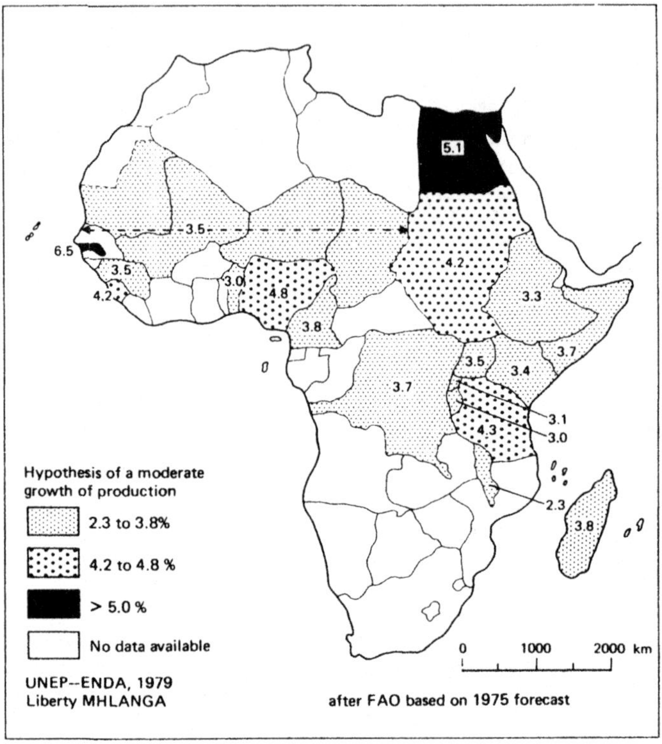

MAP 6. Agricultural growth necessary to satisfy needs in 1990.

Most often, the issues dealt with above concern quantities of food as compared with quantitative needs. It must be clearly understood, however, that even if these quantities were available, Africa's food and nutrition problem in the coming years would not be fully solved.

Food problems are in large measure caused by problems of distribution, and not simply due to the limitation of production. If these factors are to be properly interpreted, distinctions must be made between constraints and potentials at the local, regional and national levels. In examining the situation, the short-term implications should be given priority over long-term considerations.

Limitations of another kind, which are also qualitative in nature, are related to production and consumption habits. The distribution of caloric needs to be satisfied is not the same in the different zones of the African continent. While it is important that certain changes in the predominant types of food crops be foreseen over the coming years, the consumption habits and culinary practices of the various human groups will have to be taken seriously into account, for they constitute a constraint which evolves only slowly.

Other demands which should be satisfied have to do with the composition of the diet and the different elements which it should include. The same vitamins are not deficient everywhere, nor to the same degree. On the continental level, zones may be demarcated according to the actions to be taken against the most serious vitamin deficiencies (11).

As the case of a group of herders in Uganda illustrates, the composition of the diet varies with the seasonal evolution of the environment. This factor should be taken into account, for it complicates an evaluation of the situation and impedes—as it will continue to impede in the near future—an adequate distribution of nutrients to the people, the peasants in particular, and the herders even more so. There is another matter of serious concern as well. Because of a lack of sufficient transport, an inadequate distribution of purchasing power at the appropriate time of year, and a lack of education among the masses (all factors which will be difficult to change rapidly), the desirable quantity and quality of foodstuffs will not be received where they are most needed during the pre-harvest shortage. For a large part of intertropical Africa, this is the period in which agricultural work is the most laborious.

An essential factor to be taken into account in adjusting food and nutritional estimates is the unequal distribution of resources. While the most prosperous socio-economic categories tend to consume a great amount of food which they do not really need, the poorest social classes have neither direct access to, nor the means to purchase on the market, the foods which are indispensable to them.

This has a great influence on both the size and weight of children and, to a certain extent, on their performance in and out of school.[6] We have centred on the importance, for now as well as the future, of an interdisciplinary approach to problems, and a consistent evaluation of the prospects of evolution in ecological, economic and social terms. Close examination should be made, for example, of the way food problems in the various countries are reflected in the consumption patterns of the different social groups. In many cases the prices of foods and various agricultural products have been kept down to a modest level benefiting urban consumers and leading to reductions in the buying power of the rural dwellers, including their power to purchase food products. It appears that the tendency of the food supply to deteriorate will most seriously affect low-income groups, primarily those involved in agriculture, herding and transitional urban activities.

The pressing issue is to find out how, between now and the year 2000, the subsistence of the African people may be assured. It must be remembered, however, that answers which seem "technically self-evident" are often contingent on one or another of the options implicit in the scenarios of the future.

An initial solution—often compelled as a result of an emergency—is the recourse to foreign food relief. Various agencies for international food relief have been set up (in particular, FAO, WFP, Comité de Lutte contre la faim, etc.), and there is no doubt that they have helped and will continue to help in the near future to limit the effects of periods of scarcity or famine. The circumstances of continuing food assistance to Africa require close examination. Another solution, and the only one which is valid over the long term,

[6] It is true that in this case it is difficult—even using tests adapted to the cultural environment of the parents and children—to determine how much is the direct result of physical, dietary or pathological deficiencies, and how much depends on the educational level of the parents and the social milieu.

consists of setting up a more favourable relationship between men and the land they cultivate.

A lack of foreign exchange has kept certain countries from buying the quantities of cereals that they should have imported during difficult years. In the future, an increase should be anticipated in both purchases of food products, and assistance or donations. Another question which must be asked is how it will be possible in the future to reduce certain serious disadvantages of this form of assistance. An initial danger which could be accentuated over the coming years is the fact that this food assistance may no longer serve the purposes of periodic relief, but become an annual supplement to domestic production; it thus becomes a structural element of the economy of the African country.

In addition, the distribution of this assistance ought to be improved considerably. "At present the distribution centres are found in and around urban areas and do not benefit most of those who are in need and who are situated in the rural zones" (122).

One may wonder whether these gifts of food, unless there is a drastic change in the way they are allocated, will not continue to create certain obstacles: "Most food assistance is supplied by the industrialized countries as a way of disposing of their surplus. . . . In certain developing countries, it has inhibited the introduction of essential reforms and the improvement of the infrastructures necessary for agricultural development. It has also kept certain developing countries which are potential exporters from penetrating the world market" (122).

Here is another situation in which environmental considerations cannot be dissociated from the choice of the type of response to basic needs. While the food needs of the peoples living in the fringe menaced by drought and desertification may be essentially satisfied by the importation of cereals, without a rational effort in terms of both production and new territorial planning it is also certain that these risks of desertification and drought will rapidly become more pronounced.

Most of the African countries will set out upon another path.[7] As indicated by map 7, the number of inhabitants which each square kilometre of arable land will have to feed remains low; for nine tenths of the area of the continent it is less than 5 persons per hectare of arable land.

One of the vital concerns of the two coming decades will be the introduction of crops over an area of land which will be equivalent, at least, to the land which is already under cultivation.[8] African economies and societies are thus faced with a challenge: to double the cultivated land in twenty years without accelerating the destruction of the environment.

The notion of a substantial extension of cultivated land should not suggest that the areas currently under cultivation constitute a static resource. They may evolve qualitatively and quantitatively; either the land is improved or it is destroyed, through processes such as brush fires, overgrazing, the lack of an anti-salinization programme, and deforestation. Various estimates have been made of the land which could potentially be used for the extension of crops in Sub-Saharan Africa:

[7] This will depend to a great extent, it is true, on the scenario which they have chosen.

[8] It should be noted that nine-tenths of increases in African agricultural production in the 1960s were due to the extension of cultivated areas and only one-tenth due to a growth in yield.

POTENTIAL CROPLAND IN SUB-SAHARAN AFRICA

	Millions of hectares
Present arable land (FAO)	186
Possible extensions according to:	
1. FAO (IWP)	304
2. Club of Rome	423
3. Alan Strout	470
4. USA	733
5. Wageningen	653

Notes: 1974 arable area: taken from the FAO Production Yearbook. Table 1, Vol. 28-1 (FAO, 1974). Permanent crops have been included.

(1) FAO estimate for the Indicative World Plan.
(2) Calculations of Mesarovic and Pestel, 1975, "Second Report of the Club of Rome".
(3) Calculations by A. Strout (147) Ressources pour le Futur, study unpublished.
(4) US President's Science Advisory Committee (1967). The World Food Problem (Vol. II adapted from tables 6-7 and 7-9).
(5) Wageningen 1975, calculation of the maximum absolute production of food.

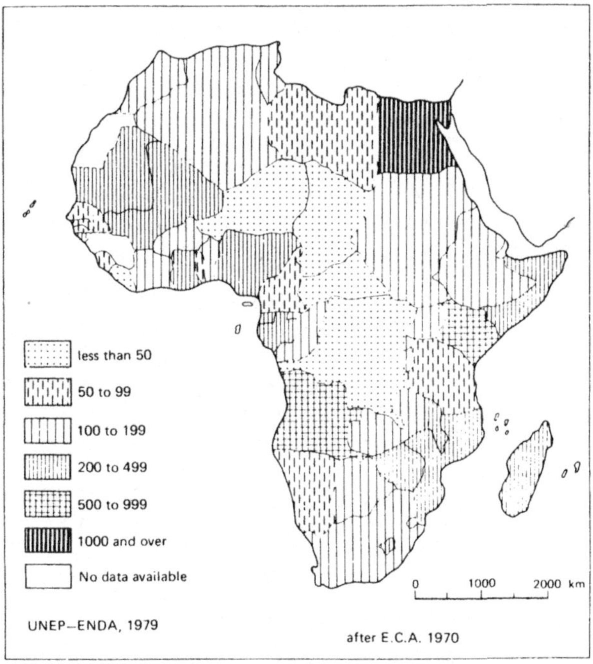

MAP 7. Human density per km² of arable land.

Theoretically there are no significant physical or technical barriers preventing the expansion of food production in Africa to keep pace with the increase in the population up to the year 2000 and beyond. The problem is in determining whether over the coming two decades this increase in production will really take place with sufficient magnitude, and in the countries and regions which need it most urgently. The economic, social and environmental costs must also be analysed.

It is possible to get an approximate idea of the new land each country will have to devote to food crops (67). An examination of the situation in each country taken individually, however, reveals limitations which make the global calculations less absolute. In a country such as Zaire, neither pressure from the people nor the economy's vast resources can justify optimism with regard to the possibilities of a large-scale exploitation of potentially arable land over the next twenty-five years (122).

At this level a series of unknown factors emerge. The first problem is whether food crops will appear sufficiently attractive (in terms of labour intensity, yields, marketing facilities and selling price) to incite the peasants to give them priority over cash crops. The second question is whether, with the resources and techniques which will be placed at their disposal over the next two decades, the farming population will be able to effectively cultivate enough land. Finally, it is important to ascertain the expected costs and who is to bear them: whether it be foreign aid, the public authorities, or the peasants themselves.

This implies a rapid examination of projected costs to regain new land and protect land which is already being cultivated from degradation and depletion.[9] The accounting must be in terms of money, various inputs, skills and days of peasant labour.

The costs of extending areas devoted to food crops can be measured in terms of quantities of land and water. Water is not available everywhere, and irrigation is not without its dangers. One should be aware of the fact that in Africa, 30% to 80% of irrigated lands are more or less threatened with salinization, depending on the context in which they are located.

Most forecasts suggest that there will be a very rapid growth in the consumption of inputs in African agriculture, a tendency which is borne out, in any event, by projections for fertilizer consumption.

The implications for the economy and for society of the allocation of substantial resources to new projects should be evaluated. Examining petroleum products, it is interesting to see what it will cost in terms of energy resources to develop the Sahel. Taking only those fertilizers judged to be indispensable, the manufacture of 500,000 tons of fertilizing units at the horizon 2000 would call for 470,000 tons of petroleum. The burden which petroleum imports of this magnitude would place on the Sahelian economies over the next two decades is known (34).

The notion that production can be augmented primarily by increasing the land areas under cultivation should be examined. Additional efforts devoted to clearing the land, an extension of pasture-land for the peasants, the risk of total depletion of certain threatened acreage by excessive cultivation and an occasional drop in yields are all aspects which should be anticipated. It has already become apparent that, despite the sevenfold increase in plantings of sesame in Kordofan in twelve years, the production of sesame has not even

[9] It is evident that even the terms of the problems will be given different weight according to the orientations chosen by the African countries during the coming decades: techniques, social organization, the respective shares of human and financial investments, etc.

doubled: yield per hectare has fallen from 0.91 to 0.21 tons—a loss of 70%.

For many villagers, the rapid increase of cultivated areas also means turning away from traditional methods of rotation. The technology used on the land which is the most seriously threatened by erosion and depletion is not always adequate. Similarly, the opening up of new pastureland to flocks, which are often too numerous, rapidly accelerates overgrazing, diminishes preferred forage species, and stimulates erosion.

The cost in terms of the difficulties faced by the rural societies or the inhabitants themselves must also be taken into account (see map 8). Thus, for the Senegal River Valley, between now and the end of the century, the radical transformation of the river's flow and the exploitation of land and livestock will lead to a dramatic transformation of the economy and society in the western Sahel. The heads of state of the countries concerned are as aware of this fact as are the planners: "the provision of water should serve to stimulate the emergence of new structures which will facilitate a subsequent acceleration of development. . . . In fact, it should be realized that the transition from traditional rain-fed and post-flood cultivation to irrigated cultivation is tantamount to a fundamental reorganization of the income structure of all of the people implicated by the water

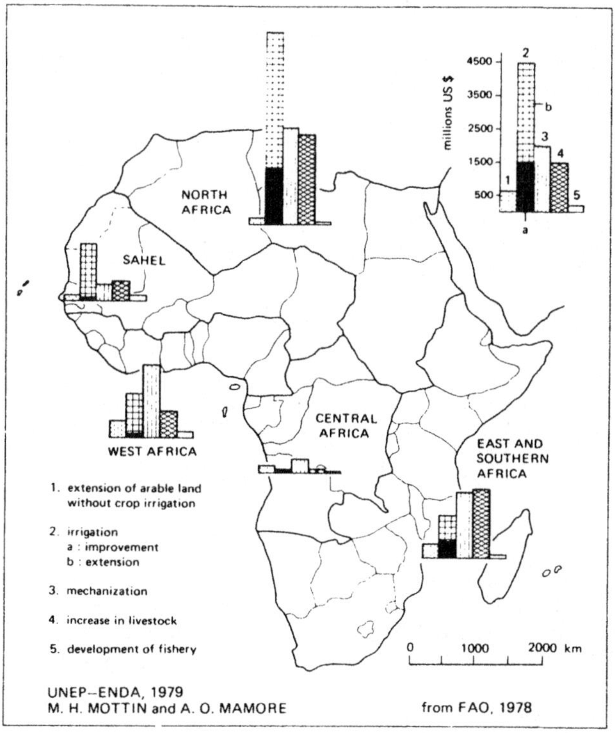

MAP 8. Investment costs necessary to attain food products on objectives 1975-90.

development, and a restructuring of the sum total of socio-economic relations between these people and the inhabitants of the neighbouring regions, whose production will also be affected."[10]

Even if they have too long, and wrongly, been considered to be the only significant factors, the importance of financial costs remains considerable. The problem of satisfying African food needs takes on another dimension when one calculates the total cost of capital investments for agricultural extensions which have been planned, and for setting up irrigation for food production in 1990. Map 8 gives only one of a number of estimates of the investment costs[11] necessary to attain food production objectives, in terms of both agriculture proper, and animal raising and fisheries, the latter being an aspect which is too often neglected or underestimated.

Fish are not prominent enough in Africa from an economic, nutritional or even a socio-psychological point of view.

There is little understanding of the potential which fishing has to offer; while the traditional fishermen are concerned with economizing shoals of fish, the same cannot be said of the latecomers to traditional fishing, most of them foreigners. Information and training provided to fishermen, with regard to this question in particular, might yield favourable results. Such an initiative would be all the more valuable in that the prospects for exploiting fish stocks are not always highly promising.

These stocks are now endangered. Foreign fleets, from the European Socialist and Mediterranean countries primarily, remove large quantities of fish from two zones situated off the African coast, along the central-eastern and south-eastern Atlantic. In the first area they take away approximately 3,500,000 tons live weight; in the second, they account for about two-thirds of the fish caught.[12]

Over the next two decades an increase in fishing off the western coast of Africa may be anticipated. For the next ten years there is hope for spectacular improvements in fishing in a few countries, such as Morocco, Ghana, Nigeria and Senegal, but much more moderate increases elsewhere (see diagram 1 and map 9).

The prospects for fishing and the role which fish may play in the diet of the continent's coastal and inland inhabitants depends on the evolution of various factors. The first has to do with fish stocks: that is, whether it will be possible to rationalize fishing, limit the depletion of certain shoals and the disappearance of certain species of fish. Another question, linked to the previous one, has to do with the catches of the foreign fleets. Not only do these fleets fish for the benefit of continents other than Africa, but they do so destructively. A third unknown factor is the rapidity with which deep sea fishing fleets belonging to the African countries can be created, on the one hand, and on the other, the rate at which small-scale local fishing can be modernized. Progress such as that achieved by Senegal in this area supports an optimistic forecast. The fourth unknown factor has to do with the extension of inland fishing to begin with, and a rational exploitation of rivers and streams, but possibly even more important, with the multiplication of small ponds where the *tilapia* in particular could thrive. Another unknown factor lies in the dissemination of techniques of smoking and conservation of fish, and the extension of marketing networks which would make fish available at a reasonable price to people living far from the coasts

[10] "Les objectifs et les grandes lignes de la stratégie du développement intégré du bassin du fleuve Sénégal", OMVS, St-Louis, May 1974, 127 pages.
[11] Estimates in US dollars at the 1975 rate.
[12] COPACE estimates for 1977-78.

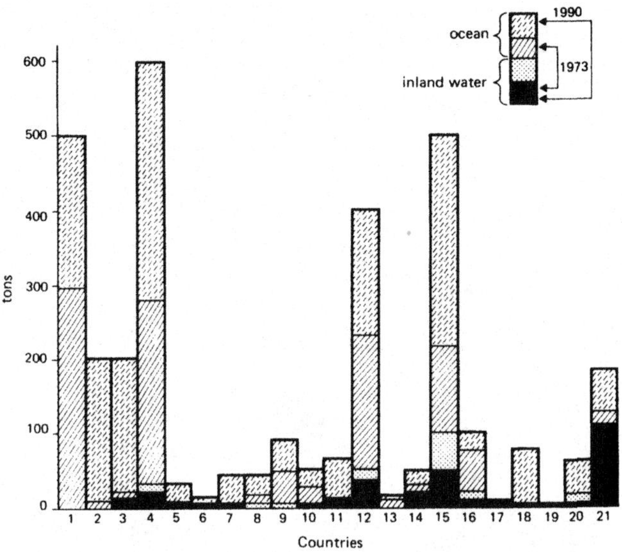

1. Morocco, 2. Ex-Spanish Sahara, 3. Mauritania, 4. Senegal, 5. The Gambia, 6. Cape Verde, 7. Guinea-Bissau, 8. Guinea, 9. Sierra Leone, 10. Liberia, 11. Ivory Coast, 12. Ghana, 13. Togo, 14. Benin, 15. Nigeria, 16. Cameroon, 17. Equatorial Guinea, 18. Gabon, 19. Sao Tome and Principe, 20. Congo, 21. Zaire.

UNEP–ENDA, 1979 after COPACE, 1976

DIAGRAM 1. Expected growth in fisheries in some African countries.

or major rivers. There is a final problem: this is the generalization of habits of fish consumption and the introduction of new dishes into the diets of certain social groups.

Thus, though it is a difficult aspect to evaluate, fishing constitutes an important element in a forecast of the answers to Africa's food needs.

Although theoretically possessing the world's greatest reserves of arable land, it will be difficult for Africa between now and the year 2000 to obtain an adequate and properly balanced diet from agriculture, animal raising and fishery, without aggravating the damages inflicted on the environment. If this challenge is not answered, it is difficult to imagine how other serious problems, such as health, might be solved.

HOW HEALTHY IS AFRICA?

An empty stomach or an imbalanced diet is a certain invitation to disease. To define health as the absence of disease would be incorrect. In fact, the healthiest nation is the one which finds the best balance between the various risks to the health of each individual and the community, and the means of controlling these risks.[13]

Making Africa healthier, often proclaimed as a fundamental objective of governments and various international organizations, would seem to call for a twofold orientation: ar-

[13] See F. Remy, 1973 and P. Campagne, 1974.

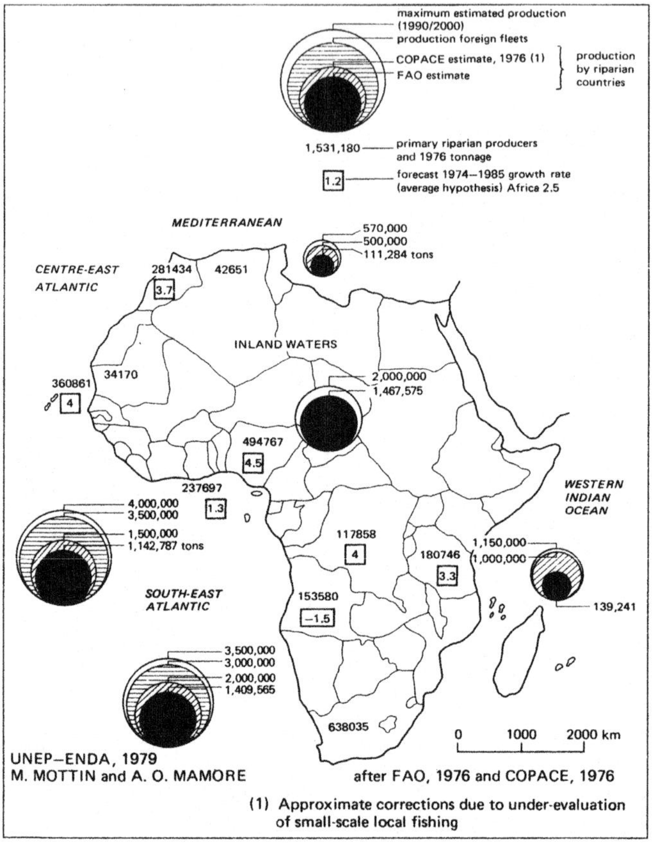

MAP 9. Prospects for fishery in Africa.

resting disease and improving the daily environment.

While it is scarcely possible to draw up a map of the present death rate in Africa, nor a diagram of its foreseeable tendencies, at the very least one can conceptualize the relative gravity of the situation. The relative scope of the efforts to be undertaken with regard to one of the combined effects of malnutrition and disease, the infant mortality rate (see map 10) is a foundation for understanding health problems in Africa. On this basis, the prospects for the evolution of the continent's worst endemic diseases[14] between now and the end of the century may be considered.

[14] According to WHO (1972) the twelve leading diseases of the African region (WHO limits) can be classified as follows: 1. malaria; 2. influenza; 3. measles; 4. trachoma; 5. whooping cough; 6. dysenteries; 7. syphilis; 8. gonococcal infections; 9. meningococcal infections; 10. infectious hepatitis; 11. leprosy; 12. bilharziasis. (Verbal report, WHO, Geneva.)

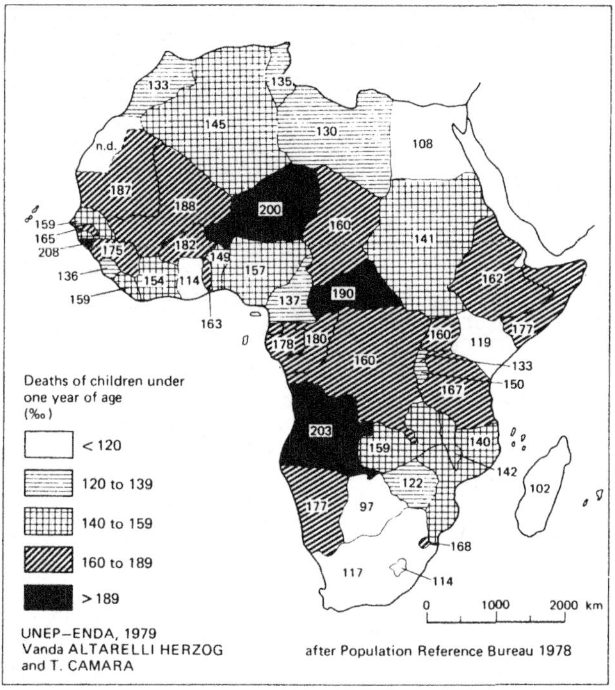

MAP 10. Rate of infant mortality.

The programme of WHO for the total eradication of malaria has been replaced by a less ambitious attempt to arrest the disease. This alteration of the objective is the result of a short budget, organizational problems and insufficient motivation of the anti-malaria units. In addition a resistance to insecticides has appeared among the vectors, and a resistance to the usual anti-malaria treatments has arisen in the parasite itself. Such resistance is likely to be a handicap to the African people over the next twenty years, making the disease more costly and complex to control, and giving reason to fear the entrenchment or the spread of this affliction.

The WHO programme for the control of onchocercosis is to be extended to new geographical areas, but the possibility of reinfestation of zones where the disease has been eradicated cannot be discounted. A few such cases have already appeared in East Africa.

Many development plans in Africa provide for the creation of dams, which would reduce the prevalence of the disease upstream, but increase it downstream. Antivectorial programmes do exist but are not yet being combined systematically with the creation of new reservoirs.

The problem of bilharziasis is related to onchocercosis, and it can be predicted with near certainty that the disease will spread as new dams and irrigation canals are built (see map 11). Similarly, it is likely that the intestinal form of the disease will take hold where it previously did not exist, resulting from migrations of workers or following ecological modifications which favour the intermediary hosts.

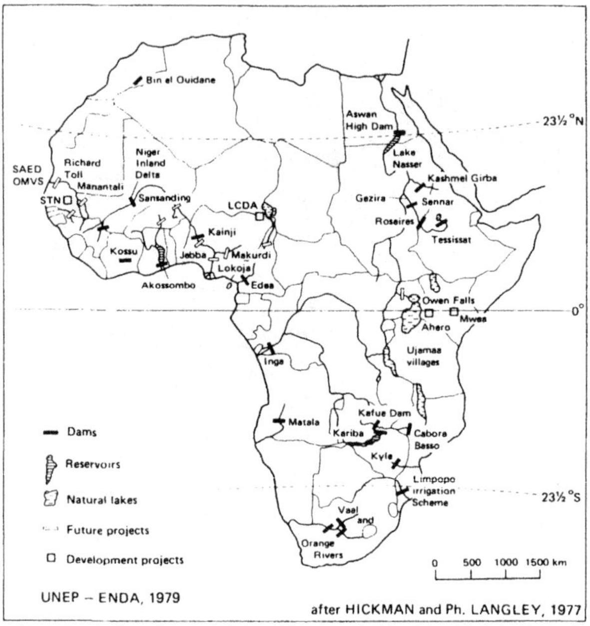

MAP 11. Dams and large-scale projects in Africa.

If bilharziasis is to be eliminated, both urination and defecation in or near the water must be stopped. Because a generalized hygiene education and the necessary installations call for considerable financial and, above all, education resources, there is little likelihood that this disease will be eradicated in the near future.

Despite the disappearance of certain centres of trypanosomiasis, the disease will undoubtedly continue to spread until it attains such proportions that its elimination will once again become a priority (see map 12).

Measles and other diseases preventable through inoculation introduce the problem of vaccinations and prophylaxis. The attitude of peasant groups and certain urban dwellers is still frequently to consider diseases as a health problem from the curative point of view only. If present efforts are continued, an improvement in vaccine coverage may be anticipated over the coming twenty years, without arriving at the current levels attained by the industrialized countries. The result of desultory vaccination programmes, as they now exist and will continue to exist, would be the appearance of sporadic epidemics of measles, polio, and whooping cough, even though their prevalence and their geographical scope would diminish.

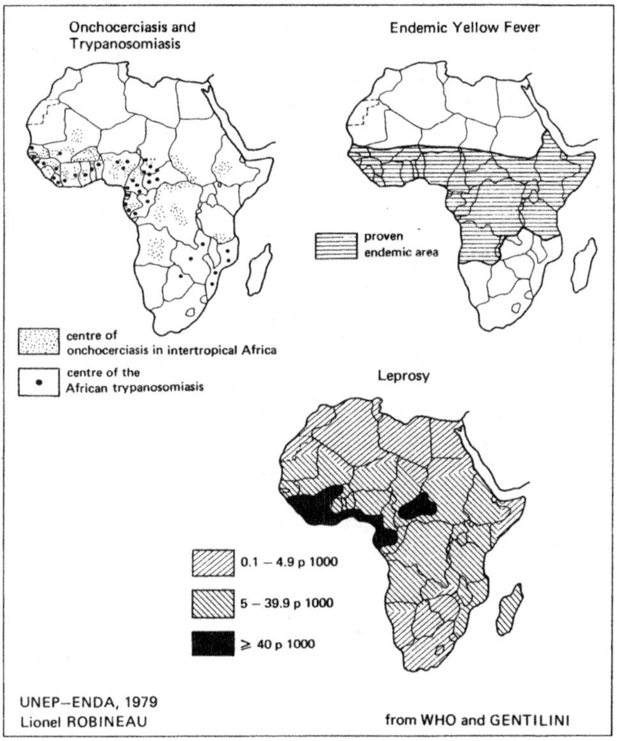

MAP 12. Zones of impact of actions to be taken against certain diseases.

With regard to yellow fever and other arboviruses, one can count on a sharp decrease in the prevalence of yellow fever in the coming years, stemming from vaccination programmes and the extended immunity which inoculation provides. This does not, however, exclude sporadic epidemics as, for example, during migration to zones which harbour an endemic disease. Total eradication cannot be expected as long as the sylvatic cycle is uncontrolled. The other arboviruses do not constitute a major threat to public health as long as they continue to be benignant, with the exception of hemorrhagic fevers (Marbourg, Ebola, Lassa) whose epidemiology is still too obscure to permit a prediction as to their future development.

Smallpox no longer seems to be a problem. The hygiene monitoring of the last known African centre of the disease has been complicated by the conflicts which are taking place there. Certain experts tend to believe that man is not the only carrier of the virus. If this were to be true, Africa would not be safe from an upsurge of smallpox once the rate of vaccination had decreased, as is now the case.

Cholera and other water-related diseases (typhoid fever, salmonellosis, shigelosis, amoebiasis, viral hepatitis, poliomyelitis) depend particularly on the environment, on personal hygiene and especially on the availability of clean drinking water. The problem is one of speeding up the creation of wells and the installation of pipes for running water so that they may exceed the growth in demand. A reasonable coverage of 80% to 90% in the rural areas might be attained by the year 2000.

Along with water-related diseases, venereal diseases threaten to increase as the cities grow. It may be hoped that by the year 2000 the problem will have been dissociated from the sense of shame and moralism which have been attached to it, thus facilitating its resolution.

It is likely that by the year 2000 the stock of Koch bacillus which have developed a resistance to one or more treatments of tuberculosis will have increased until they approximately equal the present number of resistant stocks in Europe. Improved coverage by the BCG vaccination (Bacillus Calmette-Guérin) and the improvement of early detection allow us to predict that by the end of the century tuberculosis will no longer arouse much concern, except in the case of infections of children which are manifested in adulthood.

One can count on a sharp decrease of leprosy and the total partial curing of the lepers who still number nearly 4,000,000 in Africa (see map 12). This optimism is based on the recent success of cultures of the bacillae which will make it possible to perfect the necessary vaccine, as well as the early use of Rifanipicine which totally eliminates any risk of contagion, but whose prohibitive cost at present makes it impracticable for use in large-scale prophylaxis. It is certain that leprosy will only truly disappear when underdevelopment has. It seems, furthermore, that vaccination with the BCG increases the individual's resistance to the Hansen bacillus.

There is a vaccine against types A and C of cerebrospinal meningitis; however, it is primarily group A which dominates the African meningitis "belt" (see map 13). The authors to not agree as to the efficiency of the vaccine, or as to the length of immunity

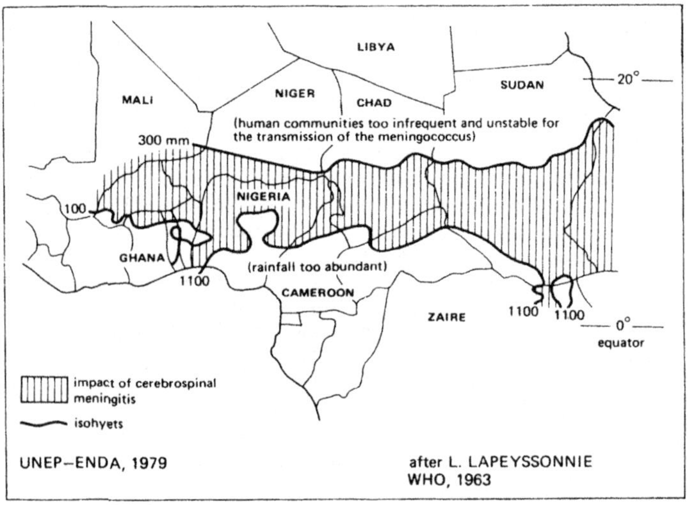

MAP 13. Zones of impact of cerebrospinal meningitis.

it provides. Large-scale improvements in housing would play a positive role, since contagion is favoured by poor ventilation.

The dreaded umbilical tetanus is beginning to diminish in areas where health education exists. It is diminishing still faster in areas where pregnant women are vaccinated and where serotherapy is practised at birth. The perfection of a low-cost immunizing anatoxin which can be administered in only two injections (possibly without a needle) allows some hope that protection against this disease will have been improved considerably by the year 2000.

Aside from primitive cancer of the liver which predominates in tropical Africa, which does not seem likely to diminish and whose aetiology is still little known, the predictable increase in industrial and urban pollution permits one to foresee a spread of cancers, which will become increasingly serious as long as pollution control is allowed to lag.

Cardiovascular afflictions are already significant because of their frequency (approximately 10% of in-hospital cases). They will continue to grow with the adoption of lifestyles which imitate Western models. Furthermore, recent studies tend to prove that high blood pressure is more frequent than generally believed, even in the rural environment.

Acculturation and the rapid expansion of the cities give reason to expect a significant increase in the cases of mental afflictions, delinquency and violence. Forms of treatment which are integrated into the cultural context and involve relatives and friends, as was traditionally the case, would undoubtedly prove truly effective in psychiatric rehabilitation. The essential problem continues to be knowing to what extent living conditions which would benefit the greatest number of people might be created in Africa during the coming two decades.

If significant progress in terms of health is sought, over the coming years an increasing interest should be shown in such factors as the quality of the water, food, housing, air, and the environment in general—evaluated in terms of their relative adequacy and the extent to which they form a harmonious and satisfying whole.

Living conditions in Africa between now and the end of the century will be significantly affected by the rate at which drinking water is made accessible to rural and urban inhabitants. It is not by chance that the countries in which at least 72% of the population do not have safe drinking water (map 14) are also those where the rate of infant mortality is equal to or greater than 160 per 1,000 (see map 10).

Water which is both more plentiful and of suitable quality can solve certain health problems as well as responding to various food and household needs. Certain governments and international agencies have made the provision of drinking water to the entire population a goal to be achieved by the end of the century. It remains questionable whether this objective is attainable, in the light of current trends.

According to WHO, there is no doubt that the situation is improving in a number of countries where the percentage of the rural population adequately supplied with water has climbed from 13 to 21% between 1970 and 1975, following an outlay of the order of $12 for each individual supplied, of which 36% was financed by foreign aid. In the urban areas the proportion of the population supplied with water has varied little in the last five years, going from 33 to 37% for private connections and 65 to 68% including public water taps. The costs are high: $67 for each citizen provided with drinking water, of which 46% has been financed by foreign aid (198).

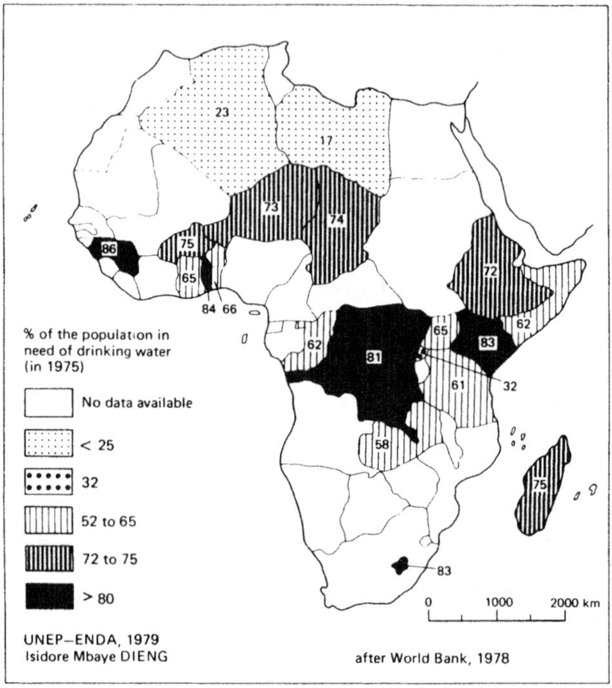

MAP 14. Proportion of population to be supplied with drinking water.

Nevertheless, the tendency which has emerged over the last few years indicates that not all the countries are doing as much as they could. With regard to the public water supply in African countries, objectives for 1980 were spelled out by WHO (see diagram 2). In the opinion of the experts consulted, these objectives have not been attained, and it would be optimistic to seek to extend the curves. To do so, for example, would yield a forecast in which drinking water in the African cities would be available to the entire population and the problem of excreta would be today eliminated.

Based on trustworthy indications, it would seem realistic to expect that the continuing growth of the major African towns and cities will create a certain backlog in both the supply of drinking water and the evacuation of excreta for that part of the population which is living in precarious housing areas.

It seems reasonable, however, to imagine that extension work and "consciousness-raising" in the rural areas will be fruitful, and that rudimentary but adequate sanitary installations will become more prevalent over the coming years. A considerable effort will be required over the coming decades. In evaluating one aspect of the problem, that of wells and the water supply in the African rural areas, "it has been estimated that the investments necessary for the coming period to attain the objectives set by WHO amount to

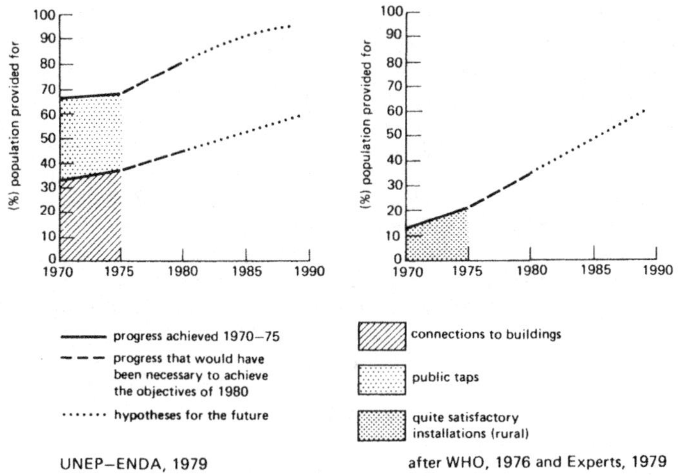

DIAGRAM 2. Forecast of providing drinking water in urban and rural areas in Africa.

approximately four times those which were made during the 1971-75 period" (198).

The elimination of excreta and the evacuation of refuse are related, if distinct, problems, and their solutions often appear slow and costly. In the urban areas it may be expected that connections to the public sewer system will be made at a rate which will not be significantly different from that of the 1970-75 period during which, in 13 African countries, the percentage of urban dwellers connected to the system went from 8 to 15% (198). During the same period the proportion of the urban population benefiting from adequate individual installations apparently increased by a fifth: possibly an unrealistically optimistic assessment of the situation. The *per capita* cost of the systems and private installations was reportedly $19, of which 13% was paid for with foreign aid.

Concurrently, during the 1970-75 period the percentage of the population having adequate installations for the evacuation of excreta went from 23% to 28%, at a unitary *per capita* cost of $7, in which foreign aid is negligible. Such a rate appears particularly slow. This leads to the question of the resources which might be used for health purposes over the coming years.

The few data which are available[15] appear too limited to permit an observation of tendencies. At most it may be observed that, in the national budgets of the African states, the share of credits earmarked for health expenditures generally is not progressing strongly. When compared with the gross domestic product of these states, it becomes apparent at least that the proportion of current health expenditures varies from 1% to 3% depending on the country. This suggests that the effort to be made over the coming years could

[15] Special research could be carried out in Africa as a whole on this specific aspect: it would be faced with the difficult task of accounting for the totality of health expenditures in the public sector—directed by several ministries (health, education, town planning . . .)—and the private sector, including associations and private individuals.

increase considerably (77). In any case, present efforts in terms of health and the future prospects of such efforts cannot be appreciated solely in monetary terms.

The probable orientation of the policies of the African states between now and the year 2000 will be towards an evolution of types of health and hygiene services in the various countries. This evolution will feature the modification of medical coverage with, perhaps, a partial correction of the serious disparities which favour the cities. Improvement in the production and distribution of medicines and products of African pharmacopea are also likely directions. Above all, a more effective prevention programme and a generalized improvement of environmental hygiene with a campaign aimed at both prevention and the education of the general public will be launched.

Choices will have to be made between the extension of European (or American) style medical care and the search for other kinds of answers to health needs. The options must be explored within the framework of the various scenarios adopted by the different states.

Hopefully the combined campaign to limit disease and improve daily living conditions will permit a significant increase in life expectancy, which should rise from the 1975 figures of 45.2 years to approximately 57 years in the year 2000. In this respect it should be noted that the efforts to be deployed vary greatly according to the country. In the Sahel, or in Angola[16] for example, considerable progress remains to be made if these countries are to come up to the African average toward the end of the century.

A DIFFERENT KIND OF SETTLEMENT

There is no doubt that a doubling of the population, the effect of migratory flows and the impact of the transformations which will take place in industry and agriculture will dramatically transform human settlements in the various zones of the African continent over the coming two decades.[17] The tendencies which have emerged indicate that there will be an increased development of productive forces in certain limited areas, a continuation of rapid urbanization and a broadening of the disparities between regions.

An extrapolation of certain aspects of the international strategies of the industrialized or Third World countries suggests that the trends which have appeared will become more pronounced. An evolution toward industrial specialization of the major regions is taking shape. The polarization of industrial activities could come about in several ways.

The industrial growth of the mineral-rich countries (central and southern Africa, and Saharan Africa) will accelerate. Eventually these areas will figure among the principal countries possessing energy resources. In addition to the extraction process, the primary processing of the extracted material will be introduced locally.

Processing activities will be clustered around the major urban centres, near export points in particular. As a result, the processes of industrial concentration, leading to rural exodus and urbanization of technology, will be reinforced. The environment will be increasingly threatened, stemming from the growing magnitude of the exploitation of its resources and the increasing demands placed on it by the urban environment (density of occupation of the land, urban development, industrialized areas).

[16] Where the life expectancy is between 38 and 39 years (132).

[17] Within the context of the present document only a very brief sketch of this problem is given. In fact, this question has already been dealt with in a number of studies carried out by ENDA, in particular, and which merit thorough consideration.

A certain increase and development of agricultural raw materials produced or processed according to the agro-industrial procedure[18] will take place in areas lacking significant mineral or energy potential: West Africa, the Sahel and East Africa (Horn of Africa). The expansion of industrial single-crop agriculture will be carried out spatially through the conquest of new land or the taking over of areas used for food production (animal raising, agriculture), according to natural climatic factors (groundnuts) or other artificial factors by means of irrigation (cotton). The major hydro-agricultural projects of this zone (Senegal, Niger, Nile) will favour these agro-industrial operations in their extension and diversification (early fruit and vegetables, sugar cane, rice). The consequences of this extension (depletion of soils, deforestation, upsetting of ecosystems) will be borne by the already fragile environment of the region, whose natural and social formations will suffer.

In some instances these products were already developed during the 1960s near the major urban centres of production (textiles) and shipping (oil mills). Where this is the case, the industrial establishments set up to process them are not likely to expand considerably. The strategies linked to the industrial policies of the industrialized and developing countries seem at present to coincide in setting up numerous industrial enterprises near the large urban centres, industrial-free zones and the extremities of the major roadways.

The major hydro-electric projects will undoubtedly contribute to the clustering of industrial development along rivers, coastal areas and the major lines of energy transport. The resulting modification of the environment will certainly be dramatic (micro-climate, vegetation, urbanization, industrialization), and lead to the appearance or aggravation of various nuisances.

The coastal areas in North Africa in particular, and the Gulf of Guinea to a lesser extent, whose communications with the major foreign centres (Europe, North America) are most direct, will attract labour-intensive processing industries. The latter will be designed to produce consumer goods for the markets of both the foreign industrialized countries and the local cities and towns. Both the urban and the marine environment will suffer the consequences of this localized concentration.

An extrapolation of the tendencies of urbanization suggests that in a number of major cities the population will multiply three or four times between now and the end of the century. It should be expected that the population will be concentrated in localities of more than 500,000 inhabitants: the relative proportion of the urban population in these cities would go from 47% in 1970 to 62% in 2000 (14).

This urbanization will be more or less pronounced depending on the zone. According to extrapolations based on an average hypothesis, it is in Arab-speaking countries and southern Africa that cities of 20,000 or more inhabitants will in fact be the most prevalent (see map 15).

Even the most optimistic hypotheses do not assume that the urban inhabitants with a low and irregular income will be able to find "modern"-type housing by the year 2000. The main difficulty over the next two decades will continue to be to ensure a minimum standard of housing and employment to the impoverished urban classes. The real question

[18] It should be noted that this kind of production may take different forms: the organization and techniques of groundnut cultivation in West Africa are different from those used in cotton growing (Sudan) or market gardening.

lies in knowing at what level urban underemployment and inadequate housing will stabilize.[19]

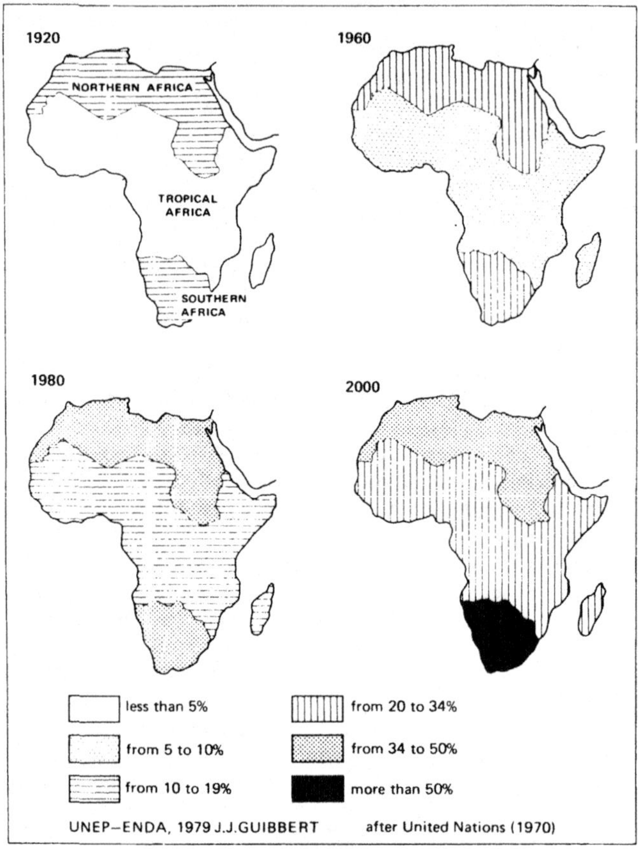

MAP 15. Urban population in Africa, 1920-2000.

It has been shown that "growth" activities will be concentrated primarily in certain environments such as the coastal zones, which are the sole gateway to the world market. The inland areas will continue to be sapped for the benefit of the cities and foreign countries. Present constraints and tendencies with regard to human settlements will limit the latitude of the policies adopted over the coming years.

At this point the following concepts should be particularly emphasized:

that the African environment is subject to serious aggression;

[19] The various orientations chosen by the states will obviously have a certain influence on the rapidity and type of urbanization and—perhaps more profoundly so—on the lot of the poorest urban classes.

that it has remarkable potentialities in the form of energy, minerals, potentials for food and other elements which could serve in the improvement of daily life (construction materials, medicinal plants, etc);

that with regard to development, both aggression and resources must be understood in relation to particular human groups which act within a circumscribed area, and in relation to the needs of these groups as well as their interests and the decisions which concern them (see diagram 3).

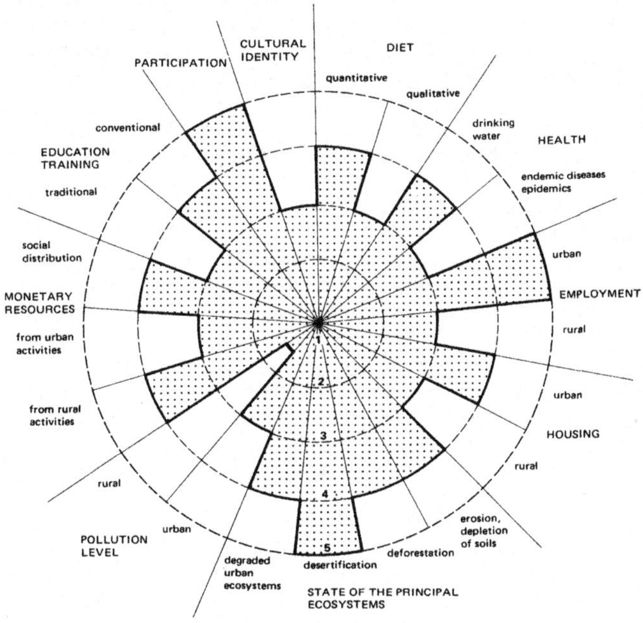

1. no problem 2. slight problem 3. problem deserving attention 4. serious problem 5. highly serious problem

UNEP–ENDA, 1979
J.BUGNICOURT and T.CAMARA

DIAGRAM 3. Importance of basic needs in Africa (an hypothesis).

The form which such aggression will take and the way these potentialities will be used are still unknown. It is questionable whether the environment can be understood and modified without reference to political decisions. Will it limit or aggravate constraints, waste or enhance the existing possibilities?

Apparently "objective" phenomena such as the desert's advance, accelerated deforestation, an abnormally high death rate, and the spread of shantytowns almost always reflect certain economic and social relations. An analysis of environmental problems inevitably implies an identification of the decisions which either complicate the situation or prevent its improvement, as well as the interests of certain groups which influence these decisions or take them directly. The drought in the Sahel would not have resulted in 100,000 deaths if priority had not been so clearly given to supplying the coastal towns with meat rather than improving the herders' living conditions, and if all of the relief provided had actually reached those for whom it was intended.

Similarly, major forest areas would not have been ravaged and denuded if a few growers had not been given free rein. The same is true for the shantytowns. In many countries they would not be so precarious if the installation of rudimentary facilities was not discouraged by land and real-estate speculation. Diseases would neither be so severe nor so often fatal if a large share of health credits were not simply channeled into the construction and operation of modern hospitals used primarily by a limited segment of the population.

It would be futile to pretend that past events will not influence the future. The major options with regard to the environment and development over the next twenty years will be choices of civilizations. For example, either a progression toward a greater integration into the present world system, or the securing of a relative margin of economic and cultural independence.

These possible forms of evolution will not take place independently of one another but will interact, thus directly determining both the framework and the content of daily life. It is options of this kind which can serve as the basis for exploring the avenues of the future.

Among the many hypotheses which may be envisaged, some can be linked due to their similar orientation in terms of development policies and the environment. Thus, a first scenario which basically involves a continuation of present tendencies is very likely to lead to an accelerated destruction of the African environments. According to a second series of hypotheses, which imply a certain autonomy of the African economies, environmental aspects would be partially integrated. In a third scenario the environment would be the point of reference for development.

The three scenarios should be examined to determine how, on the basis of present environmental data, the path of development chosen may:

(a) serve to upgrade the living conditions and environment of the African people as a whole;

(b) permit a considerable improvement in the lot of the poorest two-fifths of the African population.

The scenario is a method of imagining the future which does not mask its presuppositions and vicissitudes. Based on known factors of the present, the possible parameters of evolution are simulated by systematizing them according to the clearly stated basic hypotheses.[20] Undoubtedly the relative nature of the scenarios ought to be re-emphasized, to avoid the erroneous assumption that the African decision-makers are confined to only three possible choices. By simplifying vastly, we may say that different hypotheses have been worked out with regard to the orientation of development, the attitude toward the environment, and tendencies in the exercise of power.

[20] A. C. Decoufle writes, legitimately enough: "Any approach taken to an examination of the future is necessarily oriented by and toward normative facts. It is concerned with duration for precise reasons, and in view of determined objectives. All of these aspects should be stated before the study begins, and full clarifications should be forthcoming: these steps are absolutely necessary if the credibility of the preferred approach is to be fully guaranteed" (44).

Chapter Two

SCENARIO 1: DEPENDENT GROWTH AND ENVIRONMENTAL MISUSE

The first scenario supposes that, for the most part, present tendencies will be pursued and emerging tendencies adopted. This would mean that most of the African governments would allow themselves to be carried along with the evolutionary tide of the international economy, and by the interests of the social groups or categories which reap the greatest benefits from the present situation. If the environment hardly figures among the salient features of this scenario, it is precisely because the environment is neither a target of foreign interests (except possibly in the case of tourism) nor a concern of the wealthiest classes. When such attitudes prevail, the way in which the environment evolves is an indirect result of decisions made primarily with reference to other issues.

In this scenario, ties with the world economy and the advantages which may be derived from them continue to be thought of as the catalysts of any economic growth. The sale of raw materials abroad continues to be the major source of revenues. Hence, prospects vary according to the minerals. They are particularly promising for iron, copper, bauxite and, of course, oil. If comparative cost studies are made, despite certain concessions to be made to national interests, the multinationals and other kinds of enterprises can be induced to concentrate and even occasionally to process certain metals locally.

There are other forms of industrialization: the first, which often thrives, is the import substitution industry which manufactures "modern"-type articles and is initially aimed at a clientele which has attained a certain income level, though it may eventually find itself obliged to manufacture inexpensive articles for a larger market. In many countries this import substitution industry has already taken over many key sectors; in others, it has immense possibilities open to it.

Over the coming years certain foreign public or private sources of financing and certain local investors may show a greater inclination for the delocation industries which are developing in sectors for which the comparative costs in the already industrialized countries are no longer favourable.[21] Whether this type of enterprise will multiply depends to a great extent on the evolution of the present crisis in the dominant countries.

Another possibility is the creation or extension of industry-free zones to open the door in Africa to a number of mini Hong Kongs or mini Singapores. Their roles would be similar, though on a much smaller scale, to those played by these two cities.

There is no doubt that industrialization within the perspective of scenario 1 would result in a certain technical progress. It is also true that the repertoire of techniques available to

[21] This policy of delocation of the industrialized countries is particularly evident for the intermediary industries (steel-works, petrochemicals), and especially for common articles (such as textiles, leather, electronic components; in general these are relatively standardized articles manufactured according to techniques which have more or less stabilized, and sectors in which no important technological innovation appears imminent.

the African countries would increase. Patents would generally remain in the hands of foreign firms and, in general, manufacture would take place under their control. Technological dependence would thus increase over the period in question.[22]

This industrialization would be characterized by two important features: it would be wasteful and have a destabilizing influence. The wastefulness would stem, first of all, from the use of African raw materials without regard for their importance in a more distant future, and without any real concern for what Africa obtained in exchange. This wastefulness would also come from the mobilization of the resources available at the national level, which are usually limited, for the creation or expansion of industries which are elements of a global system. Industry would be unresponsive as a mechanism of autonomous growth. Waste would be created by a reinforcement of the attraction of the urban centres where vast reservoirs of cheap labour will continue to depress wage rates.

Another major characteristic of industrialization in the first scenario is that it generates or accentuates disparities. A prime instance of this is the disparities created between countries.[23] On the basis of resources of oil or other essential minerals, a few African countries have embarked upon a course of rapid industrial growth.

In comparison with the other African countries with which they have relations this growth has had few visible effects, or else has had a siphoning rather than a stimulating effect. There seems to be no indication that this tendency will change. Attempts to redistribute the profits derived from growth have had little success so far, and according to the assumptions of this scenario, they will make little significant headway over the coming years. Neither the Entente Council Fund nor the various agencies of West Africa, in which Nigeria holds an important place, the UDEAC (Customs Union of Central African States and Cameroon), nor the Arab League and the Organization of Petroleum Producing Countries (OPEC), have been able to help initiate activities in the poorest countries which would truly benefit their national economies. One of the hypotheses of scenario 1 is that the pattern will not be broken between now and the year 2000.

A dependent industrialization has another destabilizing side effect. Within the national territory itself, except in the case of deposits of raw materials or the case of agro-businesses in which products must be processed locally, it is the large ports and major cities, where the greatest external economies are found, which attract new enterprises. Here, once again, the hinterland supplies workers and food at a relatively low cost. It receives in exchange, through a few wages, certain extremely limited induced effects, and an often parsimonious redistribution of revenues, only very minimal benefits from the new industrial activities. Migrations of labour from the rural areas toward the in-

[22] One practice which would become prevalent, in the hypothesis in question, could probably be likened to the "turn-key" contracts—that is, those in which only one enterprise is responsible for the construction of a unit as well as for studies, orders and supplies.

[23] In fact, the most likely hypothesis appears to be that of a differential evolution, if one considers the heterogeneousness of present conditions and the improbability of an association within which the wealthiest countries would contribute to the development of the less fortunate ones. Thus, we may expect to see the effects of impoverishment appearing in certain countries which have been polarized by others where industrial growth has been rapid. Thought must be given to the relationships which might develop between these different poles on the one hand, and between these poles and the industrialized countries on the other.

dustries, between states as well as within individual countries, will probably continue or accelerate.[24]

Simultaneously a policy of agricultural growth would be pursued along three main lines. The first would involve the creation of agro-industrial concerns, in which the capital of multinational enterprises would occasionally serve to transform the environment radically. BUD-Senegal, using water obtained from the Guiers lake over 300 km away, has achieved the highest yields, while various kinds of fodder introduced from Latin America and Australia have been grown at the Aldarouch ranch in Morocco, where they feed the San Gertrudis cow, another product of cross-breeding carried out in the southern hemisphere. Enclaves of industrial production managed according to state of the art techniques could multiply near ports, airports and roads. In animal-raising areas, either private enterprises or the public authorities themselves would set up ranches situated near wells, permanent waterholes and good pastureland, for the fattening of livestock and the reconditioning of animals raised by the herders in difficult areas. There is a second tendency already beginning to emerge, which is likely to take hold: the multiplication, near centres of consumption, of small plantations which frequently belong to civil servants or urban vendors and respond directly to the demand for vegetables, fruits and poultry in the burgeoning cities. Certain technical improvements could be introduced into the small rural production sector, with the direct or indirect effect of favouring a few enterprising peasants who possess means of cultivation or transport which are more effective than those of the other villagers, and who are in a position to undertake the year-round cultivation of land which had previously been managed according to certain collective rules. As ownership is taken over by agricultural enterprises, semi-resident farmers or the best-equipped peasants, modifications in the land tenure system will be brought about with an increase in private property, regardless of legal sanctions.

The conditions necessary for the country to attain self-sufficiency in food production will hardly be fulfilled.[25] On the contrary, the increase in revenues of a certain segment of the population and the attraction of foreign models will lead to large-scale importations of foodstuffs which may include large quantities of drinks and fruits from the temperate countries.

It is likely that the policy adopted for physical organization will seek to group the people of the rural areas around the service centres so as to assist and control them more effectively. Past experience suggests that the rate of failure of such attempts will be high. In short, rural-urban relations will continue for the most part unchanged. In this scenario the

[24] In fact, there are two conflicting tendencies in the African "host" countries. One is to make increasing use of the workers from African countries who have no real civic rights and who accept work for low wages, by giving certain jobs to them rather than to the nationals and permitting a maximization of profits whether it be on the agricultural holdings or in national enterprises, or in industries in which foreign investors have a certain interest. However, the other side of the coin is the effect of the growing demagogic and xenophobic tendencies which have led in the past to the massive expulsion of workers and their families. The factor which may in the end tip the scales is the more or less willing acceptance of the dominant social classes of the most prosperous countries to share the meagre benefits of growth with the immigrant workers from the neighbouring African countries.

[25] In this scenario, food relief provided by the industrial countries will certainly be increased. In the next two decades the beneficiary countries should become increasingly sensitive to the variations in the supplies available in the industrial countries, and aware of the fact that donations are influenced by economic activity in general, and in particular, by the demand for animal products in the industrial countries themselves. The difficulty of predicting the real possibilities of the donors makes it hard for the African countries which depend on these resources to ensure the continuity of supplementary food programmes, or the implementation of certain investments using this aid.

relatively privileged social groups and categories, who are confronted with no other alternative, continue to benefit as much as possible from the surplus produced by the peasants. They are obliged to make a few concessions to the categories of peasants considered to be more "active", or a certain number of individuals of families who are part of so-called pilot activities.

The growth (disorganized, perhaps, but real) of certain industrial branches, and the prosperity of a small fraction of the agricultural sector are only comprehensible within the perspective of an urban growth. This would continue the present tendency, bringing a certain prosperity to the building industries, contractors working in this sector, and those who live off land and real-estate speculation.[26]

The likelihood, then, is that the cities will increasingly be characterized by the juxtaposition and interrelation of several urban ecosystems. In any case there will be a conspicuous imported environment, with town planning, architecture, types of housing and town management whose characteristics will largely be borrowed from the cities of the industrial countries. There will also be an infra-urban environment tied to the shantytowns, where a certain autonomy in construction and an indifference to the application of regulations cannot compensate the occupants' insecurity in terms of land ownership, the lack of equipment and the precariousness of the housing which is largely composed of materials salvaged from the refuse of the finer district. It is likely that the activities sometimes referred to as "informal" and which are in reality transitional will continue to exist in this hypothesis because they will permit the survival of a host of families whose members have been unable to find work in the "modern" sector.

There are chances that the same disproportion which currently exists between the share of public expenditures allocated to the cities and those which benefit the rural areas will continue to persist. The budget set aside for public works will be used primarily to build and repair roads in the capital, and certain highways, and the health budget largely used to operate one or two modern urban hospitals which are primarily accessible to the wealthy members of the population.

This does not mean that, in the hypothesis of a continuation of present tendencies, the governments would remain idle and, in particular, forgo their power to transform their countries' physical space. First of all, sporadic actions will be carried out in the major cities and outlying areas to defend or restore the environment. In several countries there might be a tendency toward the establishment of a special kind of zoning along socio-environmental lines. Certain well-located zones of the territory would be allocated to industrial development, whose pollution and destructiveness would officially be tolerated

[26] According to this orientation housing would probably increasingly become a commercial affair. With the exception of a few concessions to a neo-African style architecture, urban forms will continue to be copied from those of Europe and North America. Within the housing market, luxury and administrative buildings and low-cost housing for the middle classes will be the elements of a housing market for which professionals decide the norms, while the contractors, landowners and suppliers of building materials determine the prices. This means that access to adequate urban housing will continue to be difficult for a great number of people whose income lacks the stability or the level to make them eligible for any of the popular housing plans. The public authorities will continue to play a regulatory role and profess to apply a town planning code which is not really suitable in the light of local circumstances. Administrative renting on a large scale will often lead to an artificial inflation of rents. The high rates will enable the wealthiest classes to invest in speculative constructions, and pay them off in five years. For the families who occupy them, these different types of housing will occasionally provide pleasant surroundings (villas), but more often are entirely bereft of any cultural characteristics, and hardly present a fully satisfactory solution. Children are afforded precious little space. One suspects that whatever is not profitable has no place here.

on the condition that it procures profits and promotes growth. Certain districts of the cities and certain rural areas, in particular for the "week-end farmers", will benefit from services and protection which are occasionally comparable to those found in the industrial countries. Large zones of the territory could be proclaimed national parks or reserves essentially for tourist use. The "park-nationalization" of territories having no economic interest can also be anticipated with fair certainty.

It is true that this kind, or a similar kind of zoning would be justified in the light of the different problems facing the countries' leaders. In the city one can predict a numerical and possibly a proportional increase among the unemployed, the development of aspirations among the semi-educated which will be increasingly difficult to satisfy, an accelerated acculturation, the erosion of the traditional family and even the family unit, and an upsurge of delinquency and of violence in general. Where this situation continues to persist, the temptation will be great to evict the inhabitants of the popular quarters in ever greater numbers and displace them toward the periphery, setting up a sanitary cordon to protect the attractive city, and hiring sufficient policemen and public or private watchmen to supervise the population.

It is difficult to evaluate the possible consequences, in terms of collective behaviour, of the forces which are increasingly tugging at the resources and the environment, or the results of the phenomena of overcrowding, competition, and upsetting factors in individual and family life which have appeared in the large urban centres.

Within the context of these policies, tourism will continue to flourish. It should be expected that, without mustering any significant economic compensation, it will continue to play a negative role with regard to development, and propagate the effects of imitation. It has been predicted that the tourist influx from West Europe to Black Africa alone, which was 286,000 people in 1971 and 546,000 in 1976, will reach 1,500,000 in 1986 and 2,250,000 people in 1991.[27] It is possible that during the coming two decades a plan will be perfected which can be a source of significant profits: retirement plots for the aged from the industrialized countries.

The new resources derived from partial industrialization and limited modernization of agriculture, activities tied to urban growth, contributions by migrants, and profits from tourism are likely to be insufficient to satisfy the basic needs of the population. Even among those who theoretically have the means of satisfying them, the situation is in danger of evolving in such a manner that the artificially created needs will absorb a portion of the currency which would have helped cover essential needs.

In scenario 1 the possibility of satisfying the basic needs of the poor and in this way pacifying them appears to be an illusion. In fact, if we accept the hypothesis of a real growth of the economy (which, for a certain number of countries, is a likelihood), the curve of the resources available in principle to satisfy the basic needs of the majority will, over the coming years, merge with the curve of the cost of needs according to present estimates. Just when the purchasing power will be in the hands of the families whose needs are most desperate, major phenomena will be taking place which will lead to the diversion of a portion of these new resources which were believed to be available.

Without presenting the prospects of this first scenario to be worse than they are, we are forced to point out that it leads to an accelerated destruction of the natural environments in the various regions of Africa. It should be expected, in any event, that the dangers of

[27] Statistics from Union Internationale des Organismes Officiels de Tourisme (1973).

desertification will become a reality in most cases. The dense forest will be almost completely eliminated, the forests of the Sahelo-Sudanian zone, for example, will be totally sacrificed. Conditions in the shantytowns will remain as precarious as they have been, and sufficient resources will not be deployed, either to ensure an adequate diet to the poorest rural and urban dwellers, or to enable them to maintain a reasonable state of health.

A summary of the probable effects of scenario 1 is presented in the table on page 60. The serious limitations encountered in this scenario of continuing present trends are manifest. It is probable that a number of industrial countries will attempt to organize the African geo-political space to make it a politically stable zone apt to continue playing its role of supplier of raw materials to the dominant countries. It will continue to be the scene of the limitations and exploitation which dependence implies. Even where the desire to do so exists, the realization of this scenario will hardly permit popular participation in problems affecting the general public, such as education, adoption of new techniques, improvements in the diet using local resources, and protection and reconstitution of the environment (village reforestation and saving energy). It should be remembered that this first scenario is based on the principle of maximum profitability as the parameter for economic choices, and on the search for growth, with a few modifications, as the primary orientation of the action of the public authorities.

Many forecasts concerning Africa are based on the implicit hypothesis that this continent will repeat the historic experience of growth in Europe and America, destructive of both the environment and social forms which respond to the cultural values of the past. One of the decisions which will set the scene for the next twenty years involves this supposedly temporary sacrifice of the historical culture and the environment, in the interest of growth, even in its most negative industrial forms.

It should be understood that the attempt to extrapolate from an interval of time in the past to a point in the future is mechanical, and fails to express either the internal dynamics of economies and societies or external influences. The extrapolation is based on the hypothesis that the salient features of the present will remain those of tomorrow, and that there will be no change in orientation.

Rather than dealing with simple quantitative problems we are faced with a maze of qualitative problems in which the interaction of economic and social forces and the attitudes of the people have an important effect on structural variables. The fundamental question for the African environment between now and the end of the 20th century might well be: "Will history be prolonged, or is a new phase of history to begin?"

Chapter Three
SCENARIO 2: GREATER ECONOMIC AUTONOMY, MORE ATTENTION TO ENVIRONMENTAL CONSIDERATIONS

This second scenario is based on the hypothesis that Africa will have greater autonomy in growth or development, though it will still be linked to the dominant powers and that, in another sense, this situation will afford greater room for manoeuvre within which environmental considerations may find their place. It is possible that over the coming two decades a certain number of African countries, with the solidarity of other Third World countries, may impose a division of labour which would be less unequal than the one which now exists.

The elements of this scenario can already be detected in the attitude of certain countries, and the African countries as a whole. For example, negotiations which have been held by the United Nations Conference on Trade and Development, the United Nations Industrial Development Organization, the International Monetary Fund, the General Agreement on Tariffs and Trade, and the Conference for International Cooperation show that the African countries, in concert with most of the countries of the Third World, are seeking to modify certain mechanisms of the world market. In particular they are changing established patterns through the indexing of the rate of raw materials exported by Africa and the rest of the Third World, an easing of the debt in foreign exchange, a reinforcement of aid to development and greater access for their products to the market of the industrialized countries.

There is a chance that a relatively rapid industrialization of an Africa which is less dependent, or whose dependence has evolved, will take place between now and the year 2000. This more general observation can be applied to Africa: "The industrialization of the Third World is a certainty because the countries want it, because they have or will find the resources for it, and because the multinational firms and banks want to use the advantages that they offer" (38).

The similarity between this hypothetical distribution of productive tasks and other situations is that, just as in the earlier stage of growth in Africa and much like what would occur if scenario 1 were to become a reality, foreign demand would continue to be the driving force behind the economic evolution of the African countries (see in particular S. Amin (8)). However, the fact that in scenario 2 there is a relative autonomy of the African countries with regard to the dominant powers implies, unlike the first scenario, that there have been relatively significant changes in international relations.

Considering the opinions expressed by the leading officials of the different African countries and by experts, collected during the period in which the study was being prepared, and the theories which have emerged from certain models (23), it is possible to distinguish five variations within scenario 2:

There is a large-scale redistribution of oil revenues throughout the African countries.
There is increased co-operation between Third World countries.
Solid associations are formed between African states.
The revenues obtained from growth, wherever it occurs, are redistributed by the African countries, within each country.
A dynamic action is initiated in the African countries in favour of the environment.

A certain number of the advantages which Africa (and the rest of the Third World) could win back from the industrial countries will be obtained through negotiation; others through power struggles. It is likely that certain conflicts with the dominant economic powers will erupt. In general, compromises will be found, leading to a new international division of labour (see for example (8)).

Such a situation would mean a higher remuneration for raw materials produced in Africa to begin with. Several extrapolations have been made of the possible or desirable evolution of the price of raw materials exported by Africa. Among these are the forecasts proposed by W. Leontief (100) of the evolution of the price of raw materials exported by Africa:

PROPOSED FORECASTS OF THE EVOLUTION OF THE PRICE OF RAW MATERIALS EXPORTED BY AFRICA

	1980	1990	2000
Copper	101[1]	99	259
Bauxite	100	98	117
Iron ore	100	102	102
Petroleum	133	289	325
Natural gas	154	545	756

[1]Base: 1970 = 100.

It is likely that there will be an extension of systems such as that of the Lome Convention, which provides that the European Economic Community will guarantee stabilization of major imported products from 46 countries of the Third World linked by treaty to the EEC.

An assessment of the implications for Africa of present North-South relations encourages reflection on the possible future of these relations, if the policy of the industrialized countries were no longer to be dominated solely by short-term, essentially economic or defensive motivations.

The other aspect of a new international division of labour would involve most of the industrial countries relaying import restrictions to some extent to products cheaply manufactured in Africa and other parts of the Third World. In less developed nations the advantage of lower wages within a context of almost comparable productivity makes it possible to obtain a high rate of profit.

A substantial rise in the price of the raw materials exported by Africa would make it possible to step up imports of modern technologies, to make a quantum jump in industrialization and permit a rapid increase in Africa's exports to the industrial countries. Once this tendency was established it is likely that over the coming years certain conflicts of interest with the dominant forces would emerge. This would be the case, for example, if several African countries were to adopt measures similar to those of Brazil, which requires that, in the execution of equipment programmes, the rate of incorporation of national equipment attain 70%. This is despite the arguments of the World Bank which points to the increase in costs and delays in execution which such a measure will provoke.

Among the industries which develop in Africa over the coming two decades and which would export part of their production to the industrialized countries are: textile and clothing, leathers, shoes, bicycles and motorcycles, jewellery (including gold), carpentry and small-scale mechanics, as well as more complex activities such as electric construction (motors) and, for a few countries, electronics (for example, pocket calculators or hi-fi systems, various components) (40). Industrial decentralization, which appears tentatively in the preceding scenario, would come into its own.

The evolution will not be uniform: there will be jolts and jars, strides forward and slips backward. There is the possibility that, in the face of competition from the developing countries, the Western countries will react by partially or totally closing their markets for some time. Such protectionist measures could lead to difficulties of readaptation and transient crises in the African countries which would have serious consequences for certain sectors—but they could also have a more positive effect on development in the long run (40).

The rate at which this still-dependent industrialization will take place is difficult to predict. It appears probable that, in time, the lack of technology and know-how will no longer have the same importance, in particular because competition between the capitalist or socialist industrialized countries is motivating them to export the most sophisticated technology and techniques.[28]

The new orientations implied by scenario 2 will not question many of the tendencies which are already visible now and which, in the preceding scenario, are presumed to continue. Therefore we are not going to rediscuss each of the items which have already been presented. More importantly within the general framework of scenario 2, there is greater room for manoeuvre, and possibilities for increasingly making use of certain resources locally. It is valuable to calculate the extent to which different variants may come into play.

Other such variants could certainly be imagined. A comparison shows that none of these which appear here, taken individually, has either the same chance of occurring or the same potential importance in the overall scenario. Their interest lies primarily in that they provide a variety of themes for reflection.

The principal effect of a large-scale redistribution of the resources derived from petroleum would be to accentuate the general features of scenario 2 referred to above. This redistribution could be carried out in a number of ways. Once could envisage a preferential price granted by the producing countries to the African countries in general, or at least the most impoverished among them.

There is also some uncertainty as to the conditions in which this "petroleum aid" would be distributed: that is, whether it would be oriented primarily toward already established centres of industrialization, thus contributing to the creation of "poles" in a few well-

[28] The "turn-key" orders will probably be replaced by the formula of the "product control" contract, in which one supplier is responsible for seeing that the installation operates at the average rhythm foreseen. It was to this kind of revolution that President Boumedienne was referring at the meeting of Algiers (February 15-18, 1975): "It is of the utmost importance that transfers of technology to the developing countries no longer take the form simply of technical documents (delivered with a turn-key equipment), but the commitments of the promoters of the developed countries be made to apply globally to all aspects of production."

situated countries, or whether priority would be given to projects benefiting the poorest countries.[29] In any case, the magnitude of this potential redistribution would be such that it could ensure the entire range of investments necessary to put an end to the threat of famine in Africa over the next two decades.

It is difficult to judge to what extent this sub-scenario might overlap with others: how much emphasis would be given to collaboration with the rest of the Third World, or to the creation of groups of several states, to actions in favour of the poorest social classes, or to the protection and systematic enhancement of the environment. More striking than generous distribution of petroleum-derived resources, a close co-operation between developing countries could bring about a dramatic change in the conditions of evolution in Africa between now and the year 2000 (variation b of scenario 2).

Such an orientation could result in the essential modification of a number of perspectives. An Africa which is increasingly united with the rest of the Third World will have greater strength to resist the domination of the industrialized countries and to obtain trade advantages; this pertains primarily to raw materials, but may also apply to certain industries and to other areas such as tourism, for example. The African countries would greatly benefit from an intensive exchange of techniques and development experiences.[30] The extension of trade flows within the Third World would lead to a greater diversity of partners. Aid from the most advanced countries to the poorest Third World countries could take the form of technicians, technical, financial or other aid, as well as investments in various economic sectors for countries which specifically request this.

Such an orientation could be extremely interesting in terms of the environment. It would establish permanent contact between countries living in comparable ecological zones and at similar socio-economic levels, which share certain problems (endemic diseases, for example, agricultural potentials, or shantytowns) and possibilities. Fruits and vegetables which are unknown in certain parts of the intertropical zone could be introduced into these areas. The same is true for medicinal plants.

It is true that there is some scepticism with regard to the possibilities of actually arriving at a rapid intensification of South-South relations. As it has rightly been noted, the Africans, and the Third World in general, channel their efforts more toward the North-South dialogue than toward the establishment and consolidation of South-South relations (59). "An observation of the policies currently adopted by the developing countries . . . the weakness of financial and trading links between them . . . leads to the conclusion that a reorientation in favour of South-South relations will not yield major results between now and 1985" (40).

The fact that the sub-scenario is not highly probable does not make it less interesting. In addition to the economic and technical advantages which have been briefly stated, the hypotheses proposed here as a whole have a cultural interest. A greater awareness of cultural features shared by western Africa and the Caribbean, the countries and islands of the Indian Ocean, countries practising Islam (from Mauritania to the Philippines) and

[29] This raises the obvious problem of the capacity of certain countries to absorb a massive influx of investments from abroad, whether it be from petroleum producing countries or others.

[30] In a joint programme with the Department of Technical Co-operation for Development (United Nations, New York), ENDA has already begun making a modest contribution to this process with sessions in Latin America, Africa and Asia, and training programmes attached to activities carried on locally in one or another of the three continents.

countries which have assimilated Latin influences (from South America to Mozambique) would undoubtedly enable the Third World gradually to affirm an effective solidarity which would undoubtedly be of great benefit to Africa.

Another variation (c) of scenario 2 supposes the possibility of interstate associations, the consolidation of existing zones of co-operation, and the possibility in the future of enlarging them or creating others. While scenario 2 is distinctly characterized by an increase in the price of raw materials and, in general, the products exported by Africa, it is not evident that this solidarity between African states could endure beyond the specific struggle for these objectives. The next step would be to find the best openings in the market of the industrialized countries. Here one may expect to see a climate of competition developing, rather than a climate favourable to a sharing of foreign markets and the distribution of the various kinds of production which could either be placed on these markets or used to satisfy the domestic demand of various African countries, thus avoiding useless repetition and paving the way for economies of scale.

The likelihood of such associations being formed on a totally equitable basis appears limited. This variation of the scenario would come about if a few African countries, having certain advantages due to their large population (substantial production, the dynamism of certain sectors or the political will of their leaders) were to seek actively to conquer the markets of other African countries, beginning with their neighbours. The limits of the zones within which manpower is siphoned off and merchandise redistributed have already largely been defined by present migratory flows.

The creation of economic satellites would undoubtedly be achieved through the establishment of "agreements" or "common market zones". The reinforcement and broadening of existing agreements such as the West African Economic Community (CEAO), the Economic and Customs Union of Central Africa (UDEAC), the Arab League and the Comité Permanent Consultatif du Maghreb (CPCM) is likely. Satellites could evolve through new types of associations directly linked to industrial countries or groups of such countries, much like the European Economic Community/Africa-Caribbean-Pacific Group which evolved out of the Lome Convention (EEC/ACP). They could centre instead on African countries having a certain economic importance, as appears to be the case for the Economic Community of West African States (ECOWAS) (see map 16 below).

It is true that there is an "association" in southern Africa, whose evolution will undoubtedly have decisive consequences for the continent. A factor which in fact could greatly accelerate the evolution of the African continent would be a contribution of the Republic of South Africa to the development of its neighbours. The countries and territories of southern Africa as a whole constitute the most modern part of the African continent, but also the most closely tied to the world market. Industrialization, which initially reflected foreign demand, has since become increasingly responsive to the local market, unlike what has taken place in most of the African countries. The acceptance of the decolonization of Namibia and the granting of a so-called independence to several Bantustans are part of the strategy for the creation of Black satellite republics which will remain within the Rand zone, providing manpower and a market to the Republic of South Africa, while continuing to depend on it.

The environmental implications of these scenarios are particularly obvious. They would create imbalanced demographic structures consisting primarily of adolescents, women and the aged. The abandonment of certain productive activities in favour of subsistence

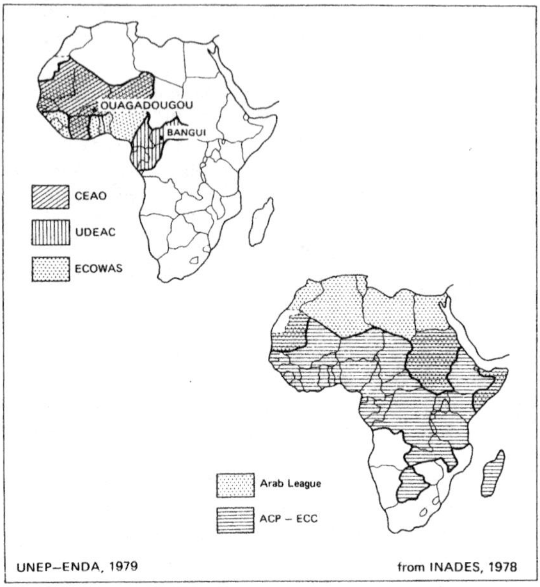

MAP 16. Constitution of enlarged economic associations.

on a migrant's income, the institution of a negative relationship between the people and arable areas, and an accelerated degradation of land, owing in particular to both rapid deforestation and increasing erosion, are going to become more pronounced.

Variation c of scenario 2 involves a great number of unknown factors which frequently crop up in the other scenarios as well.[31]

The uncertainty of the prospect of the creation or reinforcement of common markets in a number of zones of Africa stems primarily from an inability to foresee the effective consequences which will ensue. The question is whether these markets will facilitate the penetration of the multinationals, expedite the conquest of markets by the few already relatively industrialized countries or, on the contrary, permit a major redistribution which will benefit the most disadvantaged governments, countries, and regions.

Policies of a combined growth and reapportionment are the major feature of variation d of scenario 2, echoing to a certain extent the proposals of various World Bank experts.

These policies would attempt to alter the economic structure of Africa by achieving the following objectives:

A maximization of the growth of the gross domestic product through the allocation of resources to the areas which promise to be the most lucrative.

[31] This is the case for southern Africa. While it is not given further treatment below, this aspect should obviously be kept in mind.

A reorientation of investments toward the poorer groups, in terms of access to credit, equipment and education in particular.

A redistribution of revenues or a restructuring of consumption through the fiscal system and the distribution of consumer goods.

The direct transfer of advantages to the poorer groups (for example, the distribution of land).

The proposal most frequently encountered aims at attacking problems or "pockets" of extreme poverty through projects financed by foreign sources. The orientation based on basic needs fits in well here, and it is understandable that certain international organizations and industrial countries find it attractive.

This policy proceeds from the observation that the extension of the "modern" sector will not suffice to integrate the "marginal" social groups into the growth process, groups consisting primarily of peasants and a large number of urban inhabitants living in the transitional sub-system. In the urban environment, the policy aims at making certain funds available to small-scale enterprises using labour-intensive techniques, in particular the customary techniques of the transitional sub-system. In the rural areas it favours the cultivation of "new land", generally taking the form of individual appropriation and the creation of modernized holdings based on partial mechanization and the use of draught animal force.

One can clearly see the dangers of this approach which could relieve the wealthier countries of the necessity of taking more radical measures—while they provide a modicum of assistance to the developing countries. In addition this approach makes it possible for the industrialized countries to differentiate between various parts of the Third World, isolating the least developed countries and weakening the front which the dependent countries could forge in response to the policies of domination of the industrial countries.[32]

One may wonder at the probability of a "reapportionable growth" taking hold in Africa over the next two decades.

Variant d—growth, plus reapportionment—has many persuasive aspects. In particular, it is founded on the irrefutable argument that if resources are to be shared they must first have been produced. It thus finds support from both those who are interested in maximizing production and/or profits and those who stress the need to extend the benefits of development to the people as a whole.

One should be aware of the potential risks of this variant of scenario 2. The first lies in the search for a maximization of profits wherever they may be obtained—in other words, in the most lucrative sectors. Such a quest is based implicitly, and occasionally explicitly, on calculations of profitability as they are usually presented in relation to the world market, with social or environmental aspects relegated to the status of "side-effects". The world market applies its own logic in the use of profits, and within this context, the latter are not broadly available for a variety of uses, despite the leaders' best intentions. A good question is whether, given the way in which profits are accumulated, there will not be restrictions on their use.

Other difficulties would probably be created if a reapportionment were to take place. It will not be a simple matter, for example, to ensure that a reapportionment, which favours certain areas and human groups rather than others, is not a result of technocratic

[32] In fact, this policy is understandably accompanied by an increased pressure on the more conservative regimes and a more conciliatory attitude than before toward progressive experiments.

choices. The very mechanisms involved in such a redistribution would call for the running of various administrations or services whose efficiency would be difficult to ensure, and whose operation might well imply financial expenditures of different kinds.

The channels through which this redistribution would be directed to reach the beneficiaries also merit particular attention. An easy solution would be emergency allocations, or preferential prices granted, for example, through tariffs for certain goods or services. Such a solution would partially improve the food and hygiene situation, and encourage certain small urban enterprises or agricultural holdings. It would, however, be difficult to keep the beneficiaries of this distribution from coming to expect handouts as a matter of course, and eventually placing their hopes in governmental or foreign aid rather than in the potential fruits of their own endeavours. In social and environmental terms, even if a major share were devoted to this purpose, the apportionment could scarcely begin to repair the damages done as a result of the maximization of growth. In countries such as the Ivory Coast, an examination of the long-range prospects for growth/redistribution has led to a relatively precise definition of the conditions to be respected if the policy of environmental management and redistribution is not to be diverted from its objectives.

The most basic option is one of planned liberalism, founded on tacit agreement and control. Private initiative and profitability remain the major impetus behind economic activity in the field of industry. In certain countries the variation consisting of a "re-apportionable" growth may coexist with the variation in which action in favour of the environment is foreseen.

The modern fraction of society will continue to think and act as though "becoming developed" meant duplicating what others have done, importing the most modern machines, creating a few luxury neighbourhoods, having an elite with a high standard of living. The bulk of the country's resources will be siphoned off to make this modernization possible and guarantee this standard of living for a few individuals. The rest of the country and the population will have to survive as best they can. However, a certain upgrading of techniques which call for little investment and require only a moderate level of training could "console" the general public to some extent and even, in certain cases, somewhat improve their living conditions.

The limited policies of redistribution will only affect certain parts of the territory and a small percentage of the poor. Among the beneficiaries will undoubtedly be the infra-urban wards of the major cities, whose inhabitants could potentially be mobilized in a politically significant way. The implementation of site and service projects and a "supervisory assistance" directed by specialized administrative services could solve a few problems. In the rural areas certain projects, for example the setting up of small farmers with agricultural mechanization and basic tools, the provision of irrigation on a small scale, or the establishment of clinics in certain sensitive rural areas or in the home regions of the political leaders, could be carried out. It is even possible that, in order to limit the risks of a severe shortage with their socio-political consequences, to avoid an acceleration of the rural exodus and to effectively counter the predominant political power of certain urban categories, the leaders will seek the peasants' support and practise a more favourable agricultural products policy. According to one of the variations of the sub-scenario, agrarian reforms could also be carried out in certain countries where large land holdings play a significant role.

Variation e of scenario 2, the hypothesis of a vigorous action in favour of the environment, may for the time being seem hardly more than wishful thinking.

When the Stockholm Conference and the first Panafrican Conference on the Environment (Addis Ababa, 1972) were held, no one would have dared hope that six years later a majority of African countries would have their own ministries or national commissions for the environment. It also seemed improbable that a multitude of conferences and symposia would make it possible for delegates from all the countries of the continent to increase their knowledge of the basic aspects of the environment and jointly study the way things could be improved.

This is not to say that during this time the environment became the focus of the political concerns of the various African countries, but almost everywhere experience in this area increased along with an awareness of the seriousness of the problems. A certain number of effective measures were taken.

If the efforts made since 1972 are to continue, an extrapolation of trends of thinking, on the one hand, and the interest shown by top officials, on the other, suggests that any serious aggression on the environment over the coming years will be cause for circumspection on the part of decision-makers. It will kindle opposition movements among those who must suffer their consequences. The introduction of the environmental studies into school and university curricula at all levels may also increase the chances that this sub-scenario will take hold.

In scenario 2 there is to be a quantum leap in industrialization, with all of its concomitant dangers for the environment. This scenario also mandates additional resources to be placed at the disposal of the public authorities, providing them with facilities for taking measures of various kinds to protect the environment.[33]

This could mean a systematic consideration of the environmental variable in the projects and daily activities of administrative agencies. There is a high likelihood that in the year 2000, in a number of countries (particularly those having traced a path along the lines of scenario 2) environmental impact studies will be made for all major projects. The unpredictable factor, of course, is the way in which the indications provided by such studies will be applied.

Greater concern with regard to urban and industrial pollution is probable. Taking the African coast of the Mediterranean as an example, one can already predict a relatively rapid growth of industries and urban centres. Certain measures, within the framework of the Blue Plan to save the Mediterranean, are being prepared to keep the attendant dangers from being exacerbated.

Some actions will be aimed either at hastily rehousing the inhabitants of certain shanty-towns, or at improving their living conditions where they are already housed. Similarly, if the governments were really keen on doing so, the available resources would make it possible to limit a certain number of the endemic diseases referred to above, and limit the danger of epidemics. It may be assumed as well that decisive progress will be made in the improvement of the working environment.

Considerably more problematic is the question of whether these various improvements in the environment will benefit the population as a whole and constitute a global

[33] One feature common to two scenarios envisaged for the year 2000, according to a United Nations report (April 1976), is that the level of pollution in Africa will by then be 8 to 10 times greater than it is now. Nevertheless, the cost of bringing the pollution within acceptable norms would not exceed 1.2% of the gross national product of the African countries (180).

upgrading of the conditions in which they live; most likely, considerable progress will be achieved with regard to one particular aspect of the environment or another, depending on the country.

Scenario 2, in terms of both its general tendencies and its different variations, thus appears both credible and appealing. It is characterized, however, by certain ambiguities.

Not only are laws and regulations with regard to urban industrial pollution likely to spread to countries not yet having any, but their application throughout Africa will undoubtedly become more effective.

The campaign against desertification and the disappearance of the forests may also find a place in scenario 2. Since the desert's advance in particular is a theme with universal appeal, foreign funds could undoubtedly be earmarked for the tasks involved in this campaign. One might imagine, for example, that by the year 2000 an international green belt will have been established in North Africa, extending the Algerian initiative.

Africa harbours a certain number of endangered species.[34] They are endangered by the introduction of species and the disappearance of immunities or the contamination of their food. Certain species face extermination in order to protect crops, livestock, or domestic species.

It is probable that effective protection of rare animals, and possibly plant species, will rapidly become prevalent. In a number of countries the present evolution of the situation appears auspicious. This tendency could be reinforced to the extent that credits are available for the services responsible for protection, on the one hand, and for informing and teaching the public, on the other. The network of national parks and total or partial reserves will undoubtedly be extended—particularly since, within the context of scenario 2, such an extension would be linked in a number of countries to an increase in tourism. Relatively effective policies for the protection of the environment may be used by some as a way of side-stepping the need to respond to environmental concerns as a whole.

A summary of the effect which scenario 2 could be expected to have on the main environmental and development problems is outlined by the table on page 60. With the exception of a few countries, the success of scenario 2 appears uncertain and will depend to a great extent on the effects of unpredictable variables.

The first of these lies in the gap separating the intentions characterizing this scenario and the actual margin for possible manoeuvre. It is difficult to imagine how the countries which choose this "more autonomous development" within the world market system could achieve a fairly meaningful economic independence over the coming two decades. The method adopted does not appear to be adapted to the goal which has been fixed, whether the latter is construed in terms of the environment or in terms of development.

Over the coming years the ambiguity reflected in speeches or ideologies and concrete actions is likely to become increasingly serious. Continuing to encourage the pursuit of a

[34] Among the species threatened with extinction are the apes, manatees, crocodiles (2 species), the parrot (8 species), and land turtles (7 species). It should be noted that there is a convention on international trade in species of wild flora and fauna which are in danger of extinction (CITES), but a stricter application of it over the coming years would be advisable. Despite the numerous interdictions, many rural dwellers continue to live off the meat of illegally slaughtered game animals. It is foreseeable that over the coming decades it will be possible to persuade these groups to use the available animals legally and rationally.

capitalist-style growth at the same time as effective socialist-type initiatives are taken, for example, promises to be a difficult task. It is not probable that Africa will benefit from the international market while at the same time seeking to escape from it. This does not mean that dependence in certain areas cannot be reduced, but one may question whether it is realistic to expect prosperity from a situation in which decisions and needs are expressed externally, and at the same time attempt to get more internal control.

Chapter Four

SCENARIO 3: ENVIRONMENTAL DEVELOPMENT

The principal interest of environmental development is that for most African countries, the other scenarios would lead to disaster. When governments are not forced by regional wars or international crises to do so, they may voluntarily resort to environmental development. Contradictions and obstacles inherent in other present or potential approaches would lead to an increasingly uncertain development and to responses poorly adapted to environmental constraints. The prospect of "environmental development" may not be as unrealistic as it seems. Even though all the political and economic decision-makers have not been convinced, it is probable that over the next twenty years governments will begin accepting the idea that the environment/development complex constitutes a problem which should be approached from an interdisciplinary perspective. The importance of long-range perspectives in the choice of development strategies will also be accepted. It will also probably be agreed that models for national development and the implementation of plans have consequences reaching beyond national limits.

It is highly probable that the environmental dimension will be progressively included as a factor in decision-making and in economic and social progresses. Many African countries, aware that the path of development modelled on that of certain industrialized countries is not valid for them (unless the aim is to improve conditions for only an extremely small portion of the population), have had to consider other forms of development. Alternatives would make maximum use of voluntary local labour and local resources, and utilize inexpensive and preferably locally available techniques.

In defining environmental development, it is not possible to separate it from the socio-economic context in which it is proposed and from the objectives outlined in formulating it. Environmental development constitutes the form taken, at various spatial levels (villages, wards, ecological zones, regions), by an auto-centred development. That is to say, it is a development basically designed and controlled by the people concerned, and aimed to benefit the maximum number of people, based on the following:

The needs of the populations as defined by them.
A culture evolving under its own momentum.
External relations minimizing all forms of exploitation.
Local environmental resources.
Innovations and decisions made by the group itself.

Environmental development is thus, on the one hand, inconsistent with isolationism, in a state of autarchy and stagnation and, on the other hand, inconsistent with externally dependent growth which implies the imitation of foreign patterns of consumption and culture. Development in this form enables the people to invent their own solutions and ensure that their culture evolves at its own rhythm, while allowing them to integrate external contributions, technical or otherwise, according to their specific requirements. This type of approach would imply:

Giving priority to auto-centred development in all its aspects, but most importantly, in the relationship between agriculture and industry.
The strict control of resources and the search for self-sufficiency, particularly in terms of food and energy.
Taking greater account of particular regional characteristics, regional potentials and development problems.
The possibility of new forms which the international division of labour might take as well as intra-African and Third World co-operation.

This type of reorientation will, without a doubt, encounter various and extremely serious difficulties.

One of the most fundamental difficulties which would be encountered in the implementation of scenerio 3 would be that of the present role of the state in most African countries, especially with regard to the governing conceptions, the responsible administrators and their practices. Most states have inherited a more or less centralist conception of government from the colonizing countries. They must, with only a few exceptions, face the problem of having different peoples who share the same land, but speak different languages and often practise different religions. Most of the government machinery which has been set up is gradually modifying the models left by the former colonizers, but has nevertheless maintained structures and priorities which verge neither toward a decentralization of decisions, nor toward the generalized participation of the population in ecological or economic matters.

The defiance by foreign dominant powers of the generalization of environmental development formulas will constitute another threat. Certain foreign countries offering assistance may voluntarily provide aid to the inhabitants of a zone which offers no important economic or strategic advantages, in order to help them minimize their recourse to foreign subsidies. There is reason to fear that the attitude of the majority of donors would change if the new policy were to result in a limitation of exports of certain raw materials to Europe or North America. If African co-operation ended the interference of foreign economic decisions in national affairs and more or less completely blocked entry into the country of foreign consumer goods, the Western powers would change their policies. Two elements, however, could play a positive role. Support for the establishment of this new order of things could come from other countries of the Third World and from certain industrialized countries. Those African countries having chosen environmental development would continue to need certain capital goods, and while their import/export structure would undoubtedly be modified, they would still require certain know-how and tools. A large part of the market would be lost to the international economic system, but simultaneously, there would be a modification in trade structures. The African countries choosing this path would be embarking on the process of minimizing their dependency.

One of the original features of the environmental development scenario is that instead of viewing the Africa-World market relationship as static, it places primary importance on exposing the links of the system and on manipulating them.

The degree of independence of the economic system of an African country in relation to international trade can be clearly understood by evaluating the percentage its balance of foreign trade represents in the gross domestic product. This ratio reveals the relatively significant degree to which the national economies of the African countries have been integrated into the international market system. While conscious of the imperfection of this approach, one can nevertheless use it as a basis for establishing the relative dependence of various countries.[35]

In countries such as Libya, Gabon, Zaire, or Liberia, it would appear that regardless of the category in which the country falls an appraisal of their economic structures might most easily dissuade the governments from choosing endogenous development. This is the present, though not necessarily future, situation in at least one country: Mauritania. Should there be a slump in prices of raw materials or unexpected political problems impeding the flow of export goods from these countries, the authorities responsible for economic management might be obliged to modify their present positions.

For a number of countries, this would mean a complete reversal of the present economic strategy. The first requirement with which they would be faced would be, in effect, to reduce and modify their imports. Thereafter, the volume and structure of this sector could be determined by the requirements of auto-centred development, and geared to support local initiatives of both the urban and rural populations. Exports would have to be considerably modified in view of the amount of financing necessary for the importation of essential goods. Export policy would also be changed to conform to a new conception of the use of natural resources and as a result of the introduction of long-term perspectives in planning the exploitation of mineral deposits (see diagram 4).

Regardless of which African states opt for environmental development, it should be expected that they will attempt to counterbalance the consequences of their defiance of the dominant powers by establishing liaisons with neighbouring African states (especially in terms of joint development schemes)[36] and by initiating similar liaisons with the rest of the Third World.

[35] It should be pointed out that even when the degrees of extraversion are similar, the corresponding situations may nevertheless be different. First of all, among countries showing a low degree of extraversion, two different forms of economies may be distinguished. Upper Volta, Mali or Rwanda have a low level of external trade, as well as very low *per capita* gross domestic product. Their export trade is barely developed and the traditional society and economy still play an important role. The situation in Egypt or Morocco differs: the economies are more complex and better structured; the level of trade, in absolute figures, is relatively high, but proportionately, *per capita,* the value of external trade remains modest.

Among countries showing a moderate degree of extraversion, the diversity is even greater. In certain countries the absolute value of one or two exported products is high in relation to a relatively low gross product total (examples—the Ivory Coast, Senegal, and Sierra Leone). There is another category in which only one or two products dominate the external trade, but in this case they are products which command an extremely high value on the world market and constitute the bulk of their country's volume of external trade. Even though these may be countries with an important GDP, such as Algeria or Nigeria, the *per capita* value of exports remains restrained. Among countries displaying a high degree of extraversion, certain distinctions may be made: countries in which economies are based on the exportation of one or two products commanding a high market value, as in the case of Libya or Zambia; other countries, such as Zaire, where the economy presents a greater degree of complexity (83).

[36] A certain number of inter-state associations with the aim of promoting and improving regional development and planning have already been formed, for example: The Permanent Interstate Committee for Drought Control in the Sahel, founded in 1973; the Niger River Commission, created in 1964; the Organization for the Development of the Senegal River Basin, founded in 1972; the Economic Community of the Great Lakes Countries, constituted in 1976; l'Autorité de Développement Intégré de la Région du Liptako-Gourma, set up in 1971; the Lake Chad Commission, created in May 1964; la Communauté Electrique du Bénin.

It should be pointed out that part of the failure to establish certain forms of regional economic co-operation has been due to the direction in which the export trade of African countries flows. Failure has also been caused by an inadequate participation of the people concerned in the programmes undertaken. Within the context of environmental development, these agencies could, if necessary, be reoriented and restructured to contribute to the common development of the countries involved.

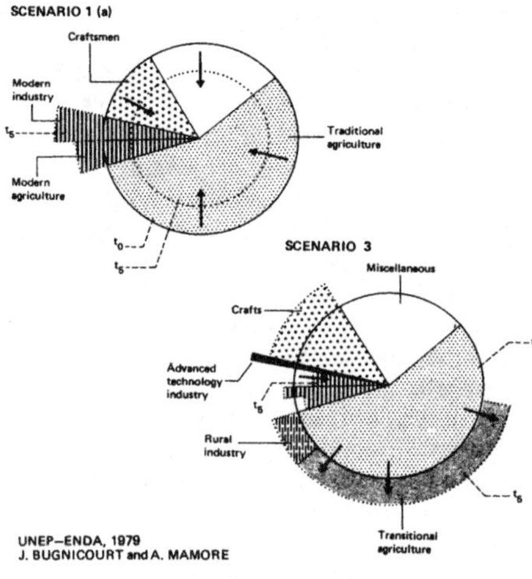

(a) Scenario 2 is more or less similar to Scenario 1

DIAGRAM 4. Environmental development implies another kind of growth.

If externally dictated needs are no longer the basic reference, a new policy of territorial development and planning can be formulated, using other types of plans[37] and a different conception of profitability.[38]

Such a strategy, founded on the response to basic needs, would be oriented toward the production of popular consumer goods such as food, clothing, housing and furniture, which encourage local production while using available resources and a high labour input.[39]

[37] The question is whether national plans will remain a mechanism of macro-economic stability or become the instruments for the application of collective choices with regard not only to production, but to the distribution of the fruits of production and to environmental development.

[38] The classic yardstick of profitability, based on comparisons drawn from the world market, is of little interest here, where the issue is one of implementing choices, putting together a long-term timetable for development based on local resources, and planning the environment.

[39] A similar approach was adopted in 1978 by Malagasy officials at the Antsirabé national seminar. The failure of "conventional strategies" having been noted, a proposal was made for a "strategy whose premise would be the satisfaction of basic needs, thus constituting a valid response to present concerns with regard to the implementation of the Charter of the Malagasy Socialist Revolution.

"This strategy allows for an economic growth which is healthy, but 'different'. Such a strategy presupposes that the composition of goods and services to be produced is determined by the people's needs. The definition of these needs cannot be arrived at solely on the basis of the market which, in the light of the existing inequalities in the distribution of monetary revenues, can only reflect the relative magnitude of these disparities. The role of the technician in this procedure is to assist the people in identifying their needs, and not to replace them. The group of people whose needs are not adequately satisfied changes as certain categories of needs are met. The evaluation of needs calls for instruments which are simple but revealing. This is the role of the indicators, nonetheless, these indicators are not neutral, any more than are the norms. The strategy of basic needs implies . . . a decentralization of planning. . . . It also implies an effective popular participation. Access to the means and the fruits of production, a major aspect of this participation, must be guaranteed by setting up of an effective organization. . . ."

The cultural aspect is a key factor in environmental development. The tendency which has emerged in the urban environment is either a limited homogeneity, crossing over ethnic lines and inhibiting the penetration of foreign cultural flows, or a cultural "pidginization", characterized by the development of one or more local cultures. In the absence of determined linguistic policies, the foreign languages of the most dynamic dominant countries can be expected to spread, along with Arabic, while certain colonial languages of the weaker countries will tend to die away.

The environmental scenario, taking into account the dynamism of certain languages (see map 17), could lead to the consolidation and expansion of particular languages within an entire group of African countries, similar to what happened in the case of Swahili, for example.

Responding to basic needs becomes the cornerstone of economic, social and environmental policy. Within the perspective of this scenario, it might be possible between now and the year 2000 to foresee the collective identification of basic needs by technicians and the people concerned, in which these needs are considered as a global whole to be evaluated with reference to local resources.

In the search for answers to basic needs within the perspective of scenario 3, technology should undoubtedly be given greater attention.

The importance of technology lies in the fact that it provides the material for the realization of economic cycles while permitting the modification of ecological cycles. Techniques are intimately linked to the structures of knowledge, the mentalities of various peoples and even to their social cosmogony. The tools cannot be isolated from their context. If they are to be used, social adjustments must be made and certain types of behaviour conditioned; certain social consequences should be expected. The transposition of technologies is a formidable enterprise. Each set of tools, processes and forms of reflection is brought into contact with a structure of knowledge and behaviours with which it may be more or less compatible.

This holds true for the most elementary technology, which is also the most widespread. Undoubtedly, combined technology is the most appropriate name for it, since in agriculture, crafts and small industry it draws simultaneously upon different sources. There are four primary sources:

Traditional African technology, closely associated with the culture of the human group involved and the environment in which they live, is an element.[40]

The technology of popular creativity, comprising inventions and do-it-yourself solutions worked out on a day-to-day basis in agriculture and industry, merits particular attention. These discoveries are worth extending, in the hope that they may encourage similar inventions.

[40] It is fairly likely that, following a period of neglect, there will be a revival of interest in traditional techniques. Such an approach presents several advantages. There is a practical advantage in that it facilitates the teaching of new technical solutions. It denotes an attitude of respect with regard to an element of the group's cultural heritage. When traditional techniques are taken into account, a greater interest is shown in the resources and constraints of the immediate environment, as well as in the underlying forms of social organization and power.

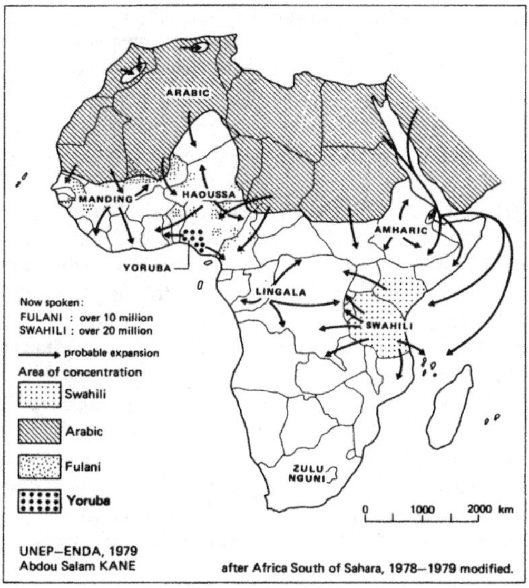

MAP 17. Prospects for expansion of some African languages.

Certain "old" technologies which are no longer used in the industrialized countries may still present a real interest for Africa. Machines which are generally simple, sturdy and inexpensive, which call for a relatively high labour input and which have been designed for small- or medium-scale production, can be useful.

Modern technologies have not been excluded from the picture; quite the contrary. The full range of possible technical responses to problems should be considered. Rather than disregarding these techniques or rejecting them wholesale, an examination should be made of their efficiency, their relative cost and their socio-cultural implications, and a selection undertaken according to local, regional and national needs.

Combined technology takes account of the different technologies mentioned above and joins them in the most appropriate way, depending on the problems to be solved in the various ecocultural zones. Naturally, within a single sequence one can resort to techniques 1 through 4, use all of them in varying proportions, or make use of only a few of them. The concept of combined technology implies that in the various African countries—and within these countries, in the various regions—choices might be made as to the kinds of technical evolution and technical combinations which would permit the most rapid development using the available means.

Cultural nationalism might also help to shape new technological orientations, particularly since many traditional techniques which have proved useful in certain areas of the continent are unknown to others. By sharing them, initial improvements might be made at a minimal expense, and the contempt which Africans have consistently been taught for their traditional techniques might be eradicated.

The criteria for selecting the optimum combination will be:

The capacity of the local group to master these techniques, to understand, implement and control them.

The environmental resources.

The prospects for evolution (eventual progress in terms of mechanization and motorization.

It should be stressed that combined technology should also imply a shared technology—in contrast to those which can only be used by highly specialized technicians. There are numerous other technologies which are likely to be adopted within the context of scenario 3, and which do not necessarily conform to the preceding definitions.

One important area will probably be energy. It should not be forgotten that Africa possesses half of the world's potential solar energy. An improvement in the processes for tapping and using this energy (a reasonable prospect) would make it possible to build a number of installations over the coming two decades. At the same time wind energy, which can be used in the Sahel and North Africa in particular, will certainly become more widely used for pumping, irrigation and milling over the coming years. These alternative forms of energy have already elicited interest from various African countries, which have been undertaking related experiments in rural and urban areas.

One aspect which will undoubtedly acquire significance within the perspective of scenario 3 concerns the techniques of recycling.

The form of industrialization featured in scenario 3 would be unlike that of the preceding scenarios. Some factories might have to be closed down and others reconverted. The creation of new factories would have as its aim directly improving agricultural productivity or upgrading the people's living conditions. In addition, this kind of industrialization would be inseparable from environmental concerns.

Concurrently, the urban transitional sub-system could be modified to satisfy the demand of rural and city dwellers for simple tools and common consumer goods. This could be a source of jobs for the unemployed.

In the environmental development scenario, the new industrial system is assured of a chance of survival only because it goes hand in hand with an equally innovative educational approach.

The vital factor in the environmental development scenario, and one which is practically non-existent in the other scenarios, is the introduction of an entirely new form of training and education. In drawing simultaneously on the old, traditional form of training, and the environment in general, this approach permits children and adults to understand their role in the environment and become familiar with the use of techniques.

In this hypothesis, the extension of a training based on the examination of techniques would permit the dissemination of elements essential to the understanding use of techniques among all the African peoples over the coming two decades. It is an invitation to extend innovations in every area. This training would be done in the national languages.

After a while, the handicap of having to resort to foreign sources for tools and processes which are not locally available could be compensated for by the large-scale dissemination of techniques. Simultaneously, this educational approach would make the people more circumspect with regard to the cultural flows which heretofore dominated the scene.

Major changes in the areas of town planning and construction would have to be foreseen, once some thought had been given to the role of the city and the ways of reducing its parasitic relationship to the rural areas. The use of public transport and means of locomotion requiring few imports and little energy input would certainly become more widespread. This in turn would lead naturally to modifications in the way the cities and their wards are laid out.

An increased use of local materials, new forms of energy (biogas or solar energy, in particular) and the services of local craftsmen would go hand in hand with the participation of the people as a whole in the construction of housing or complexes and the development of landscapes in conformity with the various cultures of the African peoples. Among the most important features of this new orientation would be a transformation in the role of the architect. Rather than building housing for paying clients or deciding upon collective norms with little or no consultation of the individuals concerned, he would essentially be responsive to the needs expressed by the people and serve as a counsellor to the inhabitants of the shantytowns, popular districts and villages in their effort to plan and build in their rural district or ward.

It should not be overlooked, however, that such an orientation demands a new attitude on the part of authorities and planners. It requires generalized training with a view to both the solution of housing problems, considered globally, and an understanding of the rudimentary techniques of construction. This would in a sense be a return to the secular practice of the African peasants who generally built their farm, mosque, or dwelling, collectively.

The perspective of environmental development does not necessarily favour one particular form of human settlement in the rural environment over another.

The orientation of Tanzania or Algeria has been to opt for clusters of villages with their own schools, services and, soon, more efficient agricultural equipment. Other countries might take a different position, requiring that major decisions with regard to the identification and organization of villages be effectively taken in concert with the people concerned, once the latter have been advised and equipped to compare the costs and advantages of the various solutions. With motorization still infrequent, it does not appear feasible to organize groups of peasants at any great distance from the fields where they work, and from which crops must be transported.

As some governments have done, the sum total of arable land could conceivably be declared to be of common utility or "public property", with the right to use it granted to the cultivators who would really put it to use. Various types of structures for production would be created, some resembling the universal co-operative model. Others would be closer to traditional communal forms or associations of age groups. The effort would serve a dual role by maintaining landscapes with a cultural value, while systematically developing the rural areas as well.

Restrictions placed on imports would make it increasingly imperative to find local solutions to problems. An effort would be made to find substitutes for fertilizers and pesticides[41] and to develop the local flora and fauna.

[41] It should be noted that a new green revolution could conceivably come about, hopefully with social consequences less serious than those of the preceding one—particularly if the ostensibly emerging tendency becomes better defined: that is, the increasing inclination to select varieties of rice, maize and wheat which can adapt to ordinary agricultural conditions and require few or no inputs from the "modern" sectors. Such a new revolution could be of interest to most of the African rural areas and make it possible to increase production without seriously increasing the peasants' dependence on the city or on foreign sources.

In every scenario, the cost of fertilizers will continue to rise as a result of increasingly costly energy expenditures, thus motivating an exploration of other solutions. The most obvious would be an increase in the production of composts. Another possible direction would be the development of varieties of cereals and grasses which can produce a proportion of their nitrogen needs by directly fixing it (from the atmosphere), production systems which increase nitrogen-fixing through bacteria or algae in the soil, or the direct application of nitrogen to the crop. There is little likelihood, however, that such solutions would become prevalent before the year 2000.

A reduction in the use of chemical pesticides will prompt a search for other techniques of combating crop enemies, based on a more thorough knowledge of ecosystems and predator-plant complexes, and on a heightened respect for the traditional practices of African peasants and herders. In addition, parasites, predators and local pathogenic agents of the crop enemies could be identified, with the immediate aim of preserving them and seeing that they are not subjected to noxious doses of pesticides. It might be possible to create the mass production of the natural opponents of crop enemies (certain entomopathogenic micro-organisms, for example) as has already been done in Chinese villages.

Along with this research on local substitute products, some effort will probably be made over the two coming decades to take greater advantage of the local flora and fauna. An African position on the conservation of flora and fauna should be defined, based on both traditional attitudes and an investigation into various possible ecological or economic uses, as well as the need to combat the damaging effects of imported technologies. This conception should take stock of prospects for production, and not simply limit itself to aesthetic or sentimental consideration as is the case elsewhere in the world. At the same time, the importance of wild animals in the cultural and social life of modern Africa should be reviewed.

Initial results of studies currently being executed in Ghana, Nigeria and Botswana, have indicated that this wildlife potential could be expanded, and that certain species could conceivably be reproduced in captivity and subsequently released, constituting an appreciable source of protein.[42] Such a prospect for the economic development of the fauna would make conservation more acceptable to those who design and apply development plans, in contrast to the position in which animals are regarded as a perpetual curiosity.

There is no reason why trees and plants could not continue to be used for the construction of housing, the production of utensils or boats, and various kinds of crafts. Similarly, it would be interesting to pursue research on plants which are currently considered to be wild grasses. In the drought-ridden zones they are already being used during the difficult periods.

Among the existing plant varieties certain species deserve special attention, such as the *Acacia albida* which stores nitrogen in its roots, has leaves and pods with a high protein content and retains its leaves during the dry season. Another interesting plant which is found in another kind of environment is the *Spirulina plantensis,* an algae found in Lake

[42] It may be anticipated that a number of states will move toward the enactment of complex protective legislation. In the end, however, the establishment of clear and workable guidelines, a judicious application of them, and an ongoing co-operation with the villagers may prove more appropriate rather than an overly rigid legal machinery which is not only complicated but impractical to apply in reality.

Chad. Traditionally, consumed by the riparian inhabitants, it contains 60% to 70% protein, and quantities of vitamins as well.

It is important to avoid a fragmentary vision of what might actually be done. One of this scenario's original features is undoubtedly the fact that, based on the felt needs of a peasant group and borrowing from its experience, development tends to be both global and progressive.

The factors which will tip the scales in favour of any of the various scenarios are the alliances which will form between the various forces which supported the national liberation movement during the decolonization stage: that is, between intellectuals, the relatively privileged social groups (linked to the political, administrative and commercial milieus), the workers, and the vast peasant majority. In this regard, the key problem is knowing the kind of relationships which will develop between the various social groups, and in particular, which of the upper-class urban groups, intellectuals or union leaders, for example, will be able to kindle the imaginations of the peasant majority.

Presented another way, the question could be whether city and industry can be expected to cease their impoverishment of the rural world: whether the urban population will discontinue its exploitation of cheap labour and accumulation of advantage through terms of exchange which are unfavourable to the peasants, forcing them to destroy their physical environment in order to subsist; or, to what extent the peasant majority will be able to give voice to its needs and aspirations, and begin to satisfy them.

Another basic question is: what, if any, approach other than environmental development can offer as many possibilities for meeting the major challenges of the African environments?

Thus, environmental development seems apt to ensure both a certain economic growth and a rational reconstruction of the setting in which people live. This does not, however, make this scenario "miraculous".

The danger to be avoided is a Manichaean presentation of the possible solutions. In fact, this would be just as risky as a moralistic conception in which good and evil are opposed. Environmental development is a scenario, nothing more and nothing less. It undoubtedly has intrinsic advantages, but also, without a doubt, presents drawbacks and risks of internal contradiction and obstacles.

COMPARISON OF SOME ASPECTS OF POSSIBLE SCENARIOS
OVER THE NEXT TWENTY YEARS

(Strong probability that the situation in some areas will worsen (Δ), remain more or less constant (C), improve (R).)

	SCENARIOS		
	1	2	3
1. Food shortage	Δ	C	R
2. Nutritional deficiencies	C	C	C-R?
3. Deforestation	Δ	C	C-R
4. Desertification	Δ	C	R
5. Erosion and soil depletion	Δ	Δ	C-R
6. Underemployment			
urban	Δ	Δ	R
rural	Δ	Δ	R
7. Difficult conditions in human settlements			
urban	C	C	C-R
rural	Δ	C	R
8. Growing risks of pollution	Δ	C	R
9. Accelerated destruction of fauna	R	R	R
10. Deterioration of landscape	Δ	Δ	R
11. Excessive exploitation of ocean's resources	Δ	Δ	C
12. Dependence and wastage with regard to energy	C	R	R
13. Under-utilization of human resources with regard to			
participation	Δ	C	R
cultural identity	Δ	C	R
14. Dangers to health	C	C-R	R
15. Possibility for 10 to 15% of the population to live according to European standards	R	R	Δ

PROVISIONAL CONCLUSION

As indicated in the introduction, this approach deliberately stresses the concept of an African entity, implying that perspectives be judiciously broadened or narrowed where necessary, and that a study be undertaken on the possible evolution of relations between Africa, the rest of the Third World and the industrial countries. At the same time prospects must be outlined, by country or by zone, with the participation of the responsible governments and interested institutions. It should be underscored that the work already effectuated only becomes truly meaningful to the extent that it is pursued and improved upon.

In the first part of this tentative glance into the future, an attempt has been made to weave together a set of forecasts as realistically as possible and in a certain sense constitute the scheme of future trends. The limitations of such forecasts are obvious, and they present certain discrepancies. Their interest lies in the fact that they do exist—meaning that they can be criticized, refined or replaced.

The second aspect of this glance into the future has been the investigation of three kinds of "eventualities", among all of those which might come to mind. These various indications should be understood in relation to the entire scheme—although it is certain that on more than one occasion they contradict or appear unrelated to it.

Obviously the scenarios as they have just been outlined cannot be taken literally. Possibly they have the virtue of indicating the multitude of general hypotheses which may come into play in determining what the African environment will be like over the coming two decades, and the extent to which future trends will be affected by a tangle of causal relationships of an economic, social and cultural nature.

Perhaps they can also indicate the value of a more exhaustive approach, of a written hypothesis as to the possible series of transformations which will come to play in Africa between now and the year 2000, and a reasonable extension of the concept of extended time "with its loose rhythms and long fibres, reconciling the fitful oscillation of events with the logic of deeper meanings" (Jacques Berque) (44).

One issue which we have kept in mind throughout this process is the room for manoeuvre which each African government has in relation to the world market, and the prevailing political conditions within and without the continent. It is difficult, for example, to predict the evolution of the international struggle for the use of resources. Given the groups of possible scenarios highlighted here, the policies of the industrial countries will undoubtedly reflect much more disparate attitudes than might at first be imagined. While the kinds of relations which they are currently establishing with the various African countries favour aspects of domination, they nonetheless include elements which appear in the three scenarios.

One of the features of the many conjectures concerning the outlook for Africa is that they are formed without consideration for the way the African rural and urban inhabitants see their own future. This study has attempted to proceed differently. So far, however, only a few results have become available. In short, it should be noted that the majority of the people, in cities and rural areas alike, feel that a "European-style" modernization of the continent will be the reality of the year 2000. Forecasts made with regard to the standard of living, on the one hand, and improvements in the setting in which people live (city, village, dwelling, urban and rural equipment) on the other, are generally extremely optimistic, and bear no relation to the possibilities of the continent as they have been envisaged in the various acceptable hypotheses.

These opinions are a reality which should be taken into account. It should be hoped that factors which have not yet become apparent will come into play to gratify these expectations.

One last remark: whether it comprises sectorial forecasts, global hypotheses or opinion surveys, the very nature of this kind of speculative exercise-outline is controversial. If it provokes contention, then it has attained its goal: that is, a discussion of the future with a view to a greater readiness to confront it, and a more accurate evaluation of the margin of voluntary choices which Africans can exercise between the various possibilities for the future of their continent.

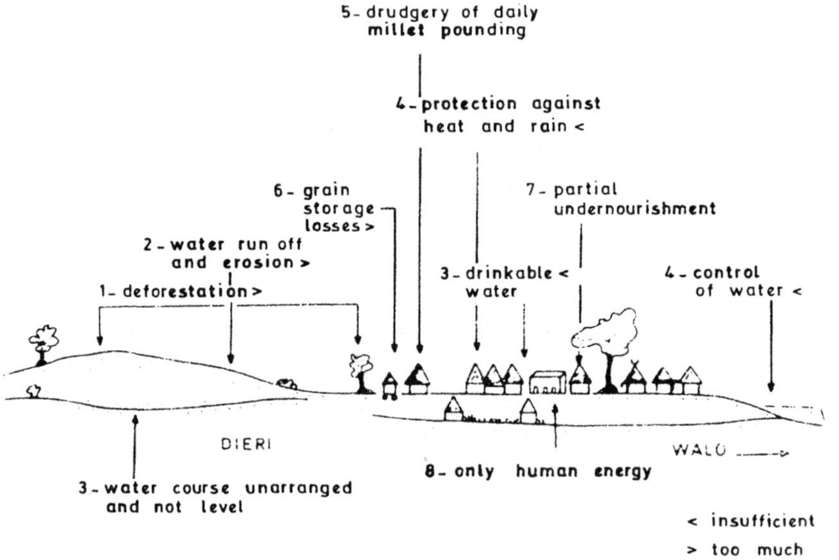

DIAGRAM 5. Principal environmental problems identified in a Toucoulou village.

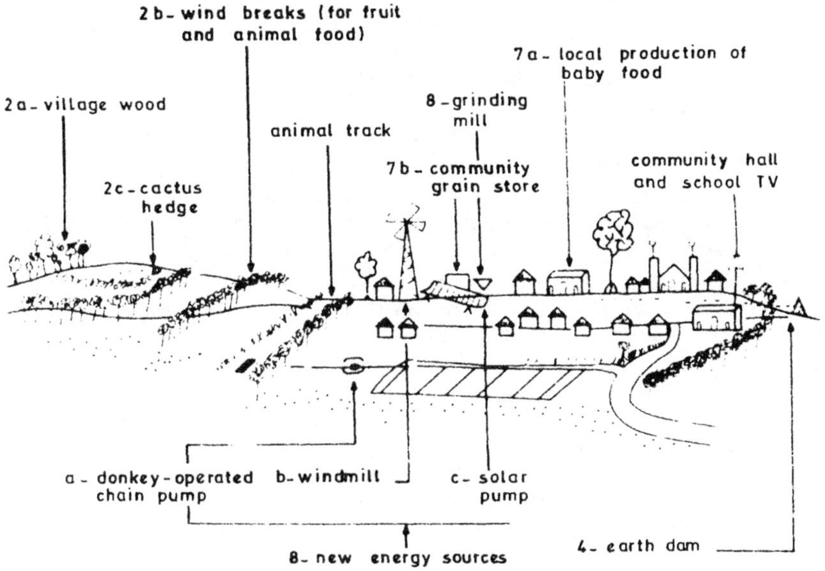

DIAGRAM 6. Possible development plan for a Toucoulou village and its surrounding district.

BIBLIOGRAPHY

(1) Abdalla, I.S., *Development in Egypt: two experiences and three scenarios,* OECD, Interfuture, 1978, 66 pp., mimeo.

(2) Aboyade, O. and Ogumpola, G.A., *Industrialization and the future economic relation in Nigeria with the industrialized countries,* OECD, Interfuture Project, University of Ife, February 1978, 81 pp., mimeo, unpublished.

(3) Adedeji, A, *Opening address delivered in Third Conference and Tenth Anniversary of the Association for the Advancement of Agricultural Sciences in Africa (AAASA),* University of Ibadan, Nigeria, 10-15 April 1978.

(4) African Environment/Environnement Africain, Quarterly Bulletin, published in English and French, Vol. I, No. 1, December 1974; Vol. I, No. 2, April 1975; Vol. I, No. 3, October 1975; Vol. II, No. 4 and Vol. III, No. 1, November 1977; Vol. III, No. 2, February 1978.

(5) African Environment/Environnement Africain, Suppléments Occasional Papers —Etudes et recherches; No. 76-1, March 1976; No. 76-5, May 1976; 76-8, July 1976; No. 76-11, November 1976; No. 77-12, January 1977; No. 77-15, January 1977; No. 77-16, April 1977; No. 77-17, July 1977; No. 77-18, August 1977; No. 77-19, December 1977; No. 77-21, December 1977; No. 78-24, May 1978; No. 78-25, May 1978; No. 78-26, June 1978.

(6) African Environment/Environnement Africain, Special issues: French, English and Spanish, No. 2, June 1978.

(7) *Agrarian Reform and Rural Development,* Paper prepared for the 10th FAO Regional Conference for Africa, Arusha, 18-29 September 1978.

(8) Amin, S., *Autocentre, autonomie collective et ordre économique international novreau,* IDEP, 1976, 25 pp., mimeo.

(9) Amin, S., *Development,* IDEP, 1976.

(10) Amin, S., *Les perspectives de l'Afrique Australe,* IDEP, 1978, 28 pp., mimeo.

(11) Arthur, D.R., *Man and his environment,* American Elsevier Publishing Company, Inc., New York, 1969.

(12) Assignment Children, *Enfants et environnement africain. Children in the African Environment,* No. 26, April-June 1974, 120 pp.

(13) Baillon, J. and Godard, O., *L'eau, limiter la croissance? Jalon pour une prospective de l'eau (l'exemple d'Alger),* in Futuribles, Autumn 1976, pp. 387-408.

(14) Bairoch, P., *Taille des villes, conditions de vie et développement économique,* Eds. de l'Ecole des Hautes Etudes en Sciences Sociales, Paris, 1977.

(15) Bairoch, P., *La croissance de la population urbaine et rurale,* Etudes Démographiques, No. 47.

(16) Balde, Kouyate and Maiga, *Stratégie de développement du bassin du fleuve Sénégal dans la problématique de l'environnement soudano-sahélien,* Session de formation et de réflexion sur les problèmes de l'environnement dans les zones arides et sub-arides, IDEP-ENDA, Niamey, February 1974.

(17) Barbedette, L., *Animation d'une zone d'extension spontanée de Douala, Cameroun,* In Malnutrition et misère urbaine, Assignment Children, No. 43, UNICEF, July-September 1978.

(18) Bene, J.G., Beall, H.W. and Cote, A., *Les arbres dans l'aménagement des terres sous les tropiques: une solution à la faim,* CRDI, 1978, 55 pp.

(19) Bertolini, G., *Lutte contre le gaspillage: politique de recyclage ou politique d'accroissement et de durée de vie des produits?* in Futuribles, No. 9, Winter, 1977, pp. 23-54.

(20) Biswas, A. K. and M. R., *Environmental considerations for increasing world food production,* UNEP, Nairobi, 1974.

(21) Biswas, A. K. and Hare, F. K., *Energy and the environment,* Comptes rendus de la Neuvième Conference Mondiale de l'Energie, Detroit, Michigan, 1974, document No. 26-3.

(22) Board on Science and Technology for International Development Commission on International Relations, National Research Council, National Academy of Sciences, *Post harvest food losses in developing countries,* Washington D.C., 1978, 206 pp.

(23) Brown, R., *Perspectives alimentaires dans le monde,* in Futuribles, No. 6, Spring, 1976, pp. 131-154.

(24) Bugnicourt, J., *Le mimétisme administratif en Afrique: obstacle majeur au développement,* in Revue Française de Science Politique, No. 6, December 1973, pp. 1239-1267.

(25) Bugnicourt, J., *Formation pour le développement, séquences technologiques d'aménagement de l'environment,* UNESCO-ENDA, Dakar, 3ème édition, 1976, 44 pp. (publié en version préliminaire abrégée par la revue Tiers Monde), No. 54, April, June 1973, pp. 369-401.

(26) Bugnicourt, J., *Pour l'aménagement des campagnes africaines une formation-action,* in Tiers Monde, April-June 1973, Paris, pp. 369-401.

(27) Bugnicourt, J., *Which urban alternative for Africa?* Vol. II, African Environment, No. 3, ENDA, Dakar, 1976, pp. 3-20.

(28) Bugnicourt, J., Cisse, B.M. and Sall, A. *et al., Environnement et Aménagement en Afrique,* revue Tiers Monde, Tome XIX, No. 73, January-March 1978, 216 pp.

(29) Campagne, P., in Assignment Children, UNICEF, No. 26, 1974.

(30) Charbonnier, D., *Prospects for fisheries in the Mediterranean.* Ambio, Vol. 6, No. 6, 1977, pp. 374-376.

(31) Clark University, *Final report to analyze environmental issues and trends in Eastern and Southern Africa,* Program for International Development, Worcester, April 1977, 65 pp.

(32) Clergerie, B., *Projet pour une recherche: stratégies éducatives et modèles étrangers,* in *Recherche pédagogique et culture,* No. 21, February 1975, pp. 24-29.

(33) Club du Sahel—CILSS, *Stratégie et programme de lutte contre la sécheresse et de développement dans le Sahel: Bilan des travaux accomplis en 1977-78, Esquisse d'un programme de travail pour 1979,* Amsterdam, 1978, 47 pp.

(34) Club du Sahel—CILSS, *L'énergie dans la stratégie de développement du Sahel: situation, perspectives, recommendations.* Paris-Ouagadougou, October 1978, 155 pp.

(35) Club du Sahel—CILSS, *Projet de compte rendu de la troisième conférence du Club du Sahel. Amsterdam, 21 au 23 novembre 1978,* Paris-Ouagadougou, January 1979, s.p.s. (4 annexes, mimeo).

(36) Club du Sahel—OECD, *Strategy and Programme for Drought Control and Development in the Sahel,* May 1977, 127 pp.

(37) Comeliau, C., *Choix de développement en Afrique et politiques de pays industrialisés,* OCDE/INTERFUTURE—Conference sur l'Afrique et la problématique du futur, Dakar, July 1977, 29 pp., mimeo.

(38) COPACE/PNUD/FAO, *Bulletin d'Information,* Dakar.

(39) CSTD, *Obstacles au développement des zones arides et semi-arides,* Rapport d'un groupe ad hoc intersecrétariat, conformément à la résolution 1898 (LVII) de l'ECOSOC, Comité de la science et de la technique au service du développement, February 1975, 33 pp.

(40) Commissariat Général du Plan, *Rapport du groupe chargé d'étudier des économies du Tiers Monde et l'appareil productif français,* January 1978, 2 vol., s.p.c., mimeo.

(41) Conférence Internationale sur la Schistosomiasie, *Country profile: Egypt.*

(42) Cvjetanovic, B., Grab, B. and Uemura, K., *Dynamics of acute bacterial diseases: epidemiological models and their application in public health,* WHO, Geneva, 1978, 143 pp.

(43) Dag Hammarskjöld Foundation, *What now? Another development, Development dialogue,* special issue, 1975, 128 pp.

(44) Decoufle, A.C., *La prospective,* PUF, 1972, 124 pp.

(45) Demailly, S., *L'Afrique noire et la division internationale du travail—Multinationales et délocalisation industrielle,* IDEP, June 1977.

(46) Direction Générale du Plan, *Les options fondamentales pour la planification socialiste,* Antanarivo, 1978, 60 pp.

(47) Dowidar, M., *Alternative strategies, African development and environment,* UNEP, Nairobi, 1977 (draft restricted), 137 pp.

(48) Dumont, R., *L'Afrique Noire est mal partie,* ed. du Seuil, Paris, 1962.

(49) Dumont, R., *Paysans écrasés, terres massacrées,* R. Laffont, Paris, 1969, 359 pp.

(50) ECA and UNRIST, *Report on a unified approach to development analysis and planning, note by Secretary General,* E/CN.5/519, 5 December 1974, (2) *Application of a unified approach to development analysis and planning under African conditions,* E/CN.14/CAP.6/4, 30 September 1976.

(51) Eckholm, E.P., *The other energy crisis, firewood,* Worldwatch Paper, September 1975.

(52) Eckholm, E.P., *La terre sans arbres,* Paris, Laffont, 1977.

(53) ENDA, *Quels futurs pour l'environnement africain? (Document de travail),* Conférence sur l'Afrique et la problématique du futur, UNITAR-IDEP, July 1977, Dakar, 46 pp.

(54) Erny, P., *L'enfant et son milieu en Afrique Noire,* Paris, Payot, 1972, 310 pp.

(55) Everett, G.C., *Développement et planification halieutiques dans la région du COPACE: vue d'ensemble,* COPACE, FAO, PNUD, Rome, October 1976, 71 pp.

(56) Everett, G.C., *Les pêches dans l'Atlantique Centre-Est—Les pêches africaines du Nord Ouest. Problèmes d'aménagement et de développement,* COPACE, FAO, PNUD, Dakar, February 1978.

(57) Everett, G.V., *Eastern Central Atlantic Fisheries—The fisheries of Guinea, Sierra Leone and Liberia: Observations on their management and development,* COPACE, FAO/PNUD, Dakar, May 1978.

(58) Fahri, A., *L'industrie africaine et la division internationale du travail,* IDEP, Dakar, July 1977, 123 pp., mimeo.

(59) Fahri, A., *Stratégies du nord et stratégies du sud: quelques hypothèses pour l'Afrique—version préliminaire,* IDEP, July 1977, 62 pp., mimeo.

(60) Fawzy, Mansour, *Les pays en voie de développement entre les théories de la croissance économique et la théorie du développement socio-économique,* IDEP, Dakar, April 1975, 14 pp.

(61) FAO, *Perspective study of agricultural development for the Arab Republic of Egypt, land and water development and use,* Land and Water Development Division, Rome, April 1973.

(62) FAO, *Perspective study of agricultural development for the Democratic Republic of the Sudan, land and water development and use,* Land and Water Development Division, Rome, April 1973

(63) FAO, *Etude prospective pour le développement agricole des pays de la zone sahélienne (1975-1990),* Vol. I: rapport principal; Vol. 2: annexes; Rome, 1976, 230 pp.

(64) FAO, *Annuaire statistique des pêches,* 1976, Vol. 42, Rome.

(65) FAO, *Evaluation des ressources halieutiques de l'Atlantique Centre Est,* Rome, February 1976.

(66) FAO, *State of natural resources and the human environment for food and agriculture,* SOFA, Rome, 1977.

(67) FAO, *Agriculture: towards 2000,* Technical Working Group of the ACC Task Force on Long Term Development Objectives, 24-28 July 1978, Geneva, Global Perspective Studies Unit, Economic and Social Policy Department, Rome, s.p.c., June 1978.

(68) Francis, G., *Prospective sur l'écodéveloppement, le développement national et les politiques de coopération internationale,* Ottawa, 1976, 21 pp.

(69) Gakou, M.L., *Performances de la production agricole et évolution de l'agriculture africaine de 1950 à 1973,* IDEP, Dakar, November 1976, mimeo.

(70) Gakou, M.L., *L'Afrique et la problématique du futur—Eléments de réflexion sur la stratégie des besoins essentiels dans le monde rural—Etude partielle,* IDEP, June 1978, 73 pp., mimeo.

(71) Gakou, M.L., *Quelques éléments de réflexion sur les problèmes pratiques de la technologie dans l'agriculture africaine,* IDEP, July 1978, 16 pp., mimeo.

(72) Galal, S., *Pertes de sol dans la vallée du Nil,* in Uniterra, Nairobi, August 1977, pp. 4-5.

(73) Gentilini, M. *et al., Médecine tropicale,* Flammarion Médecine-Sciences, 2e edition, 1977, 561 pp.

(74) GERDAT, *La situation alimentaire dans les pays en développement associés,* Ministère de la Coopération, Paris, September 1978, 162 pp.

(75) Goux, C., Deleusse, C. and Demailly, S., *Prospective Sénégal 2000—Sénégal: trois scénarios pour l'an 2000,* Laboratoire de conjonctive et prospective, October 1977.

(76) Gupta, B., *The changing role of the major multinational oil firms,* in Multinational Firms in Africa, C. Widstrand (ed), IDEP and SIAS, Upsala, Sweden.

(77) Hamid, G.M. and Gakou, M.L., *Evolution économique et sociale de l'Afrique (1950-1975),* IDEP, 1976-1977, 10 fascicules, mimeo.

(78) Hamid, M.G., *The exploitation of mineral resources in Africa,* Seminar on technology and industrialization in Africa, Algeria, January 1978.

(79) Hamid, M.G., *The copper industry and its perspectives in Africa,* IDEP, July 1978.

(80) Helmer, R., *Pollutants from land-based sources in the Mediterranean,* AMBIO, Vol. 6, No. 6, 1977, pp. 312-316.

(81) Henry, P.M., *Culture et développement,* OCDE, Paris, November 1976, 10 pp. (provisional paper).

(82) Henry, P.M. *The Mediterranean: a threatened microcosm,* AMBIO, Vol. 6, No. 6, 1977, pp. 300-307.

(83) Herrera, A.O. *et al, Catastrophe or new society? A Latin American world model,* Fundacion Bariloche, Buenos Aires, Argentina & CRDI, Ottawa, 1976, 108 pp.

(84) ICRISAT, *Annual report 1976-1977,* Hyderabad, India, 239 pp.

(85) IEDES, *Particularités des échanges commerciaux et différenciations spatiales en pays sous-développés,* Paris, 1975, 67 pp.

(86) IFPRI, *Meeting food needs in the developing world: the location and magnitude of the task in the next decade,* Washington, February 1976, 64 pp.

(87) IFPRI, *Agricultural investment and input requirements for increasing food production in low-income countries,* working paper, provisional, June 1978, 134 pp.

(88) ILO, *Basic needs, objectives and policies in long-term development planning,* 1977.

(89) INADES, *Afrique Economique,* Abidjan, 1978, 182 pp.

(90) Johnson, V.O.T., *Problem of the environment and environmental education in Africa,* African regional seminar on environmental education, UNESCO, Brazzaville, September 1976.

(91) Kane, F., *L'écolier dakarois et sa ville,* Dakar, 1974, 35 pp., mimeo.

(92) Ki-Zerbo, J., *L'éducation traditionnelle,* ENJE-YUGA.17, ENDA, Dakar, 1975, 16 pp., mimeo.

(93) L'exploitation des océans, l'économie de mer, éditions PUF, 1977.

(94) Labonne, M., *Etude prospective pour le développement agricole des pays de la zone sahélienne (1975-1990), méthodologie et procédures opératoires,* FAO, Rome, 1976, 105 pp.

(95) Langley, P., *Choix technologiques pour l'environnement construit: un essai prospectif sur la cartélisation des produits de bâtiment,* ENDA, Dakar, 1976, 34 pp.

(96) Laya, D., *A l'écoute des paysans et des éleveurs du Sahel,* in Environnement Africain, Vol. 1, No. 2, ENDA, Dakar, April 1975, pp. 53-101.

(97) Le Moal, Y, *Eléments pour l'intégration des questions spécifiques à la sidérurgie dans une réflexion sur l'avenir de l'Afrique dans le DIT,* GRESE-IDEP, July 1977.

(98) Le Lourd, P. *Oil pollution in the Mediterranean Sea,* AMBIO, Vol. 6, No. 6, 1977, pp. 317-320.

(99) Lecomte, B., *Essai: le technicien et l'environnemental rural,* séminaire DECDAK, UNESCO-ENDA, Dakar, January 1975, 8 pp.

(100) Leontief, W. *et al., L'avenir de l'économie mondiale,* ONU, 1976.

(101) Lerner, J., *Vers une nouvelle stratégie de développement urbain,* Nouvelles de l'Ecodéveloppement, No 5, June 1978, pp. 31-39.

(102) Lund (University of), *Science, technology, and basic human needs,* Report of the Lund Conference, Research Policy Program, Sweden, June 1977, 12 pp.

(103) McNamara, R. S., *Address to the Board of Governors, 1976,* World Bank, Manila, 1976.

(104) Makhijani, A., *Energy policy for the rural Third World,* International Institute for Environment and Development, London, 1976, 58 pp.

(105) Mazingira, *Satisfaire les besoins fondamentaux,* Mazingira No. 7, 1978.

(106) Meadows, D.H., Meadows, D.L., Randers, J. and Behrens, W.W., (MIT), *Rapport sur les limites de la croissance,* Delaunay, J., *Enquête sur le Club de Rome,* FAYARD, 1972, 314 pp.

(107) Mhlanga, L., *Workshop on rural environment and development planning in southern Africa—Report of the workshop,* ENDA, Dakar, July 1976, mimeo.

(108) Ministère du Plan, *Côte d'Ivoire 2000: groupe de réflexion prospective, rapport final en conclusion dex travaux,* in Plan Quinquennal de Développement Economique, Social et Culturel 1976-1980, Vol. I, Abidjan, 1977, pp. 26-58.

(109) Morawetz, D., *Twenty-five years of economic development 1950 to 1975,* World Bank, Washington, 1977.

(110) Morrock, R., *South Africa's Bantustans: illusion and reality,* IDEP, Lusaka, February 1974, 28 pp.

(111) Moumouni, A., *Retour aux langues et cultures nationales,* Perspectives, Vol. 1, 1975, pp. 68-75.

(112) Moyana, J.K., *The political economy of sanctions and implications for future economic policy,* IDEP, Dakar, July 1977, 52 pp.

(113) National Council for Research, *Sudan's desert encroachment control and rehabilitation programme,* Khartum, 1976.

(114) National Academy of Sciences, *Energy for rural development, renewable resources and alternative technologies for developing countries,* Washington, 1976, 307 pp.

(115) Norman, C., *Soft technologies, hard choices,* Worldwatch Paper 21, June 1978, 48 pp.

(116) Nyerere, J., *Indépendance et Education,* Clé Yaoundé, éd., 1972.

(117) OECD, *Study of trends in world demand of major agricultural commodities,* Paris, 1976, 349 pp.

(118) OECD, *Rapport du groupe d'animation sur la lutte contre les ennemis des cultures vivrières des petits agriculteurs dans les pays en développement,* Groupe de planification de l'OCDE sur la science et la technologie pour les pays en développement, OCDE, Paris, March 1977, 72 pp.

(119) OECD, *Environnement physique,* Résultats intermédiaires du projet de recherche INTERFUTURS, Paris, May 1977, 54 pp., mimeo.

(120) OECD, *Energie.* Résultats intermédiaires du projet de recherche INTERFUTURS OCDE, August 1977, 143 pp., mimeo, unpublished.

(121) OECD, *Matières premières,* Résultats intermédiaires du projet de recherche INTERFUTURS, OCDE, January 1978, 139 pp., mimeo.

(122) OECD, *La nutrition et l'agriculture,* Résultats intermédiaires du projet de re-

cherche, INTERFUTURS, OCDE, March 1978, 115 pp., mimeo, unpublished.

(123) OECD, *Development prospects in developing countries,* Intermediate results of the INTERFUTURS research project, Paris, 1978, 156 pp., mimeo, unpublished.

(124) OECD, *Patterns of evolution of developing countries,* Intermediate results of the INTERFUTURS research project, February 1978, 136 pp., mimeo.

(125) Ominde, S.H. and Ejiogu, C.N. *et al., Population growth and economic development in Africa,* Keinemann, London/Nairobi/Ibadan, in association with the Population Council, New York, 1972, 421 pp.

(126) Organisation Mondiale de la Santé, Cinquième rapport sur la situation sanitaire dans le monde 1969-1972, Actes officiels de l'OMS, Genève, 1975, 334 pp.

(127) OMVS, *Etude sociologique du bassin du fleuve Sénégal—Rapport intérimaire (fin de phase d'observation qualitative),* Direction de la Planification et de la Coordination, Haut Commissariat de l'OMVS, Dakar, June 1978, mimeo.

(128) Ornauer, H., Wiberg, H., Sicinski, A. and Galtung, J., *Images of the world in the year 2000—A comparative ten nation study,* Humanities Press, Atlantic Highlands, N.J., 1978, 729 pp.

(129) Osterberg, C. and Keckes, S., *The state of pollution of the Mediterranean Sea,* AMBIO, Vol. 6, No. 6, 1977, pp. 321-326.

(130) Patel, S.J., *Energy policies and collective self-reliance of the Third World,* UNCTAD, Geneva, March 1978, 17 pp., mimeo.

(131) Poore, D., *Conservation des écosystèmes,* IUCN, 1978, 29 pp.

(132) Population Reference Bureau, *World population growth and response,* Washington, D.C., 1976.

(133) Poulsen, G., *Man and tree in tropical Africa,* Three essays on the role of trees in the African environment, IDRC, 1978, 31 pp.

(134) Rapp, A., Le Houerou, H.N. and Lundholm, B. *et al., Peut-on arrêter l'extension des déserts? Une étude plus particulièrement axée sur l'Afrique,* Rapport publié en coopération par les programmes d'environnement de l'ONU (UNEP) et par le Secrétariat suédois pour l'Ecologie Internationale (SIES), Ecological Bulletins, No. 24, 248 pp.

(135) *Regional economic Atlas, Africa,* Oxford, Clarendon Press, 1967, 164 pp.

(136) Remy, F., *Assignment Children,* UNICEF, No. 23, 1973.

(137) Rweyemamu, *Underdevelopment and industrialization in Tanzania,* Nairobi, Oxford University Press, 1973.

(138) Sale, J.B., *Recommendations arising from a documentation study of the non commercial utilization of wild plants and animals in Africa,* usable for FUTAF, International Union for Conservation of Nature and Natural Resources, Morges, Switzerland, May 1978.

(139) Sachs, I., *Environnement et styles de développement,* in Environnement Africain, Vol. 1, No. 1, December 1974, pp. 9-34.

(140) Sachs, I., *Regard hérétique sur deux modèles mondiaux,* Mazingira, No. 1, 1977, pp. 6-11.

(141) Sankama, A., *Inga et l'avenir industriel au Zaire,* IDEP, July 1977.

(142) Sasson, A., *Développement et environnement, faits et perspectives dans les pays industrialisés et en voie de développement,* Mouton, Paris-La Haye, 1974, 423 pp.

(143) Savane, L., *Le futur de l'Afrique: perspectives de population,* UNITAR-IDEP, Dakar, July 1977, 23 pp.

(144) Secrétariat d'Etat aux Affaires Etrangères et Counseil de l'Entente, *Approvisionnement en viandes de l'Afrique centre ouest,* SEDES, Paris 1968, 87 pp.

(145) SEDES, *Les systèmes ruraux sahéliens,* Paris, 439 pp.

(146) SEMA, *La réalisation d'une voie ferrée transahélienne et le développement du Sahel, Etude préliminaire des effets que pourrait avoir sur le développement à long terme des Etats sahéliens de l'Afrique de l'Ouest, la réalisation d'un axe de transport lourd Est-Ouest.* (METRA-INTERNATIONAL) Marketing et Modèles de Décision, Paris, November 1977, 99 pp.

(147) Strout, A., 1975, *Ressources pour le futur,* unpublished study.

(148) Subrahmanyam, D.V. and Pineo, C.S., *Community water supply and excreta disposal situation in the developing countries: a commentary,* WHO, Geneva, 1975, 41 pp.

(149) Tandon, Y., *Role of transnational corporations and future trends in southern Africa,* IDEP, Dakar, July 1977, 17 pp.

(150) Thacher, P.S., *The Mediterranean action plan,* AMBIO, Vol. 6, No. 6, 1977, pp. 308-311.

(151) Thomas, L.V., Under the direction of, *Prospective du développement en Afrique Noire. Un scénario: le Sénégal,* PUF, Paris, 1978.

(152) Thomas, C.Y., *Industrialization and the transformation of Africa, an alternative strategy to multinational companies expansion,* A paper presented to the Conference on Multinational Companies organised by IDEP, Dakar, 1974.

(153) Tinbergen, J., *Comment faut-il étudier l'avenir?* Centrum voor Ontwikkelingsprogrammering, Erasmus Universiteit Rotterdam, s.i.d., 27 pp.

(154) Tomaselli, R., *The degradation of the Mediterranean Maquis,* AMBIO, Vol. 6, No. 6, 1977, pp. 356-362.

(155) Technology Relay (ENDA), *Exchange network on technology, Amélioration de l'eau et de l'assainissement d'un quartier—propositions pratiques; A self-reliant environment? The possibility of a technological alternative; Les images éducatives: comment les concevoir; The role of biogas in rural development,* December 1977, mimeo series.

(156) Union Internationale pour la Conservation de la Nature, *Liste des Nations Unies des Parcs Nationaux et Réserves Analogues,* Morges, Switzerland, 1967.

(157) United Nations, *Concise report on the world population situation in 1970-75 and its long range implications,* Pop. studies, 56, 1974, ST/ESA/SER A/56.

(158) United Nations, 1970, *Growth of world urban and rural population,* New York.

(159) United Nations, Committee for Development Planning, eleventh session, 7-18 April 1975, *Environmental impacts on the growth and structure of the world economy,* E/AC.54/L. 76, 2 April 1975, p. 11.

(160) United Nations (General Assembly), twenty-ninth session, A/C/2/292, 1 November.

(161) United Nations, *Population prospects as assessed in 1968,* Population Studies, No. 53, New York, 1973.

(162) United Nations, *Revised framework of principles for the implementation of the new international economic order in Africa,* 1976-1981-1986, L/CN.14/ECO/90, Rev.3.

(163) United Nations, *Pour une croissance économique commune, un dialogue nord-sud effectif et un développement plus autonome,* Conférence des Nations Unies sur la Coopération Technique entre Pays en Développement, Buenos Aires, Argentine, 30 August - 12 September 1978, Case studies on TCDC.

(164) UNCTAD, *Transfer of technology; its implications for development of the New International Economic Order in Africa,* New York, 1978, 54 pp.

(165) UNECA, Deuxième rénunion africaine sur l'énergie, Nations Unies, Conseil Economique et Social, CEA, Accra, 1-12, March 1976.

(166) (a) *Situation actuelle et perspectives en matière d'utilisation de l'énergie solaire en Afrique,* document établi par l'UNESCO, mimeo, 10 pp.

(167) (b) *Les ressources géothermiques potentielles de l'Afrique—Recommandations en vue de leur mise en valeur,* document établi par le Centre pour les ressources naturelles, l'énergie et les transports du siège des Nations Unies, mimeo, 12 pp.

(168) (c) *Effets sur l'environnement de la production et de l'utilisation d'énergie en Afrique,* document établi par le PNUE, mimeo, 29 pp.

(169) (d) *Développement des raffineries de pétrole en Afrique, situation actuelle et perspectives d'avenir,* document préparé par le Secrétariat de l'ONUDI, 37 pp. plus annexes, mimeo.

(170) (e) *Le bois en tant que source d'énergie,* document établi par le Secrétariat de la FAO, 3 pp., mimeo.

(171) (f) *Formation de personnel technique hautement qualifié dans le domaine de l'énergie en Afrique,* document établi par l'UNITAR, 11 pp., mimeo.

(172) (g) *Développement et perspectives de l'énergie électrique en Afrique,* ronéo, 27 pp.

(173) (h) *Evaluation de la situation énergétique en Afrique et de ses perspectives futures,* 27 pp., mimeo.

(174) UNEP, *Plan d'action pour la Méditerranée et acte final de la conférence de plénipotentiaires des Etats côtiers de la région méditerranéenne sur la protection de la mer Méditerranée,* New York, 1978, 56 pp.

(175) UNEP, Working group of the Committee for Development Planning, New York, 11-15 December 1978. *Environmental considerations in the formulation of the new international development strategy—discussion note submitted by the secretariat of the United Nations Environment Programme,* 15 pp., mimeo.

(176) UNEP, *Etat de l'environnement 1974,* Nairobi, 1974, 20 pp.

(177) UNEP, *Etat de l'environnement 1975,* Nairobi, 1975, 21 pp.

(178) UNEP, *Etat de l'environnement 1976,* Nairobi, 1976, 22 pp.

(179) UNEP, *Etat de l'environnement: sélection de sujets—1977,* New York, 1977, 16 pp.

(180) UNEP, *Alternative approaches to development that have been proposed within and outside the UN system,* provisional working document, 1977, 100 pp.

(181) UNEP, *Transnational project on management of livestock and rangelands to combat desertification in the sudano-sahel regions (SOLAR),* A/CONF.74/26, United Nations Conference on desertification, Nairobi, 1977, 69 pp.

(182) UNEP, *Transnational green belt in North Africa,* A/CONF. 74/25, United Nations conference on desertification, Nairobi, 1977, 54 pp.

(183) UNEP, *The state of the environment: ten years after Stockholm,* project plan (final draft), State of the Environment and Special Assignments Units, Nairobi, September 1978, 90 pp.

(184) UNESCO, *Colloque de Belgrade sur l'éducation relative à l'environnement.* Paris, 1976, 51 pp.

(185) UNESCO-UNEP-FAO, *Tropical forest ecosystems,* 1979, 683 pp.

(186) UNIDO, *Directory of fertilizer production facilities,* Part I—Africa, New York, 1970, 271 pp.

(187) UNIDO, *Le développement industriel et l'environnement,* Rapport du Secrétariat de l'ONUDI pour la Conférence des Nations Unies sur l'Environnement, Stockholm, 1972.

(188) UNIDO, *Rapport fu colloque sur la population, les ressources et l'environnement,* Bucarest, Roumanie (E/CONF.60/CBP/3), 25 March 1974.

(189) UNIDO, *L'industrie des engrais dans les PVD, situation actuelle, perspectives de développement et coopération international,* Ve session du Conseil du développement industriel, Vienne, December 1974 and January 1975, February 1975, 33 pp.

(190) UNIDO, *Draft worldwide study of the iron and steel industry. 1975-2000,* prepared by the International Center for Industrial Studies, ONUDI, December 1976, 214 pp., mimeo.

(191) UNIDO, *Draft worldwide study of the fertilizer industry,* prepared by the International Center for Industrial Studies, UNIDO, Vienna, 1976, 267 pp.

(192) UNIDO, *Draft worldwide study of the petrochemical industry 1975-2000,* prepared by the International Center for Industrial Studies, UNIDO, Vienna, 1978, 255 pp.

(193) UNIDO, *Technology for solar energy utilization,* Development and Transfer of Technology Series No. 5, UNIDO, New York, 1978, 155 pp.

(194) UNITAR-IDEP, *Compte rendu de la réunion Futur de l'Afrique,* Dakar, December 1977, 24 pp.

(195) US Bureau of Mines, *Mineral Year book,* Vols. I, II, & III, 1973, 1974, 1975.

(196) US Bureau of Mines, *Mineral facts and problems,* bicentennial edition, 1976.

(197) WEBBE, G., *Control of Schistosomiasis in Ethiopia, Sudan and East and West African countries,* Max J. Miller ed., proceedings of a symposium on the future of schistosomiasis control; Tulane University, New Orleans, 1972.

(198) WHO, *World Health Statistics Report—Rapport de statistiques sanitaires mondiales,* OMS, Vol. 29, No. 10, Geneva, 1976, pp. 544-632.

(199) WHO, *World Health Statistics Annual; annuaire de statistiques sanitaires mondiale.* Vol. II, maladies infectieuses: cas et décès, Geneva, 1978, 204 pp.

(200) WHO, *Formulating strategies for health for all by the year 2000: guiding principles and essential issues,* Geneva, January 1979, 59 pp.

(201) Williams, M.J., *Stratégies pour le développement du Sahel: le Club des Amis du Sahel,* OCDE, November 1976, pp. 195-216.

(202) Williams, M.J., *Le Sahel horizon 2000: une stratégie globale de développement.* In L'Observateur de l'OCDE, No. 89, November 1977, pp. 27 *et sq.*

(203) *Williams, M.J., Coopération pour le développement: efforts et politiques poursuivis par les membres du Comité d'Aide au Développement,* OCDE, Paris, November 1978, 296 pp.

(204) The World Bank, *World development report 1978,* Washington D.C., August 1978, 121 pp.

(205) Wright, W.H., *Geographical distribution of schistosomes and their intermediate hosts,* in Ansari, Epidemiology and Control of Schistosomiasis.

This is an abridged version of a report prepared for the seminar on "Alternative Patterns of Development and Life Styles for the African Region" jointly sponsored by the United Nations Environment Programme and the United Nations Economic Commission for Africa at Addis Ababa (5-9 March 1979),

Among the persons who contributed to the realization of the initial report were:
A. BOARE, K. BOUBANE, J. BUGNICOURT, B. COULIBALY, A. DIA, B. DIENG, I. DIENG, N.A. DIOUF, M. FALL, J.J. GUIBBERT, A.S. KANE, F. KOUME, L. MHLANGA, M.H. MOTTIN, E.S. NDIAYE, L. ROBINEAU, A. SALL, V. SKROBISCH, R. WINSHALL.

Maps and diagrams were drawn by T. CAMARA and A.O. MAMORE.

V. ALTARELLI-HERZOG was responsible for the preparation of this abbreviated version of that report.

Printed in Dunstable, United Kingdom

91221472R00050